Alterations in the Neuronal Cytoskeleton in Alzheimer Disease

ADVANCES IN BEHAVIORAL BIOLOGY

Recent Volumes in this Series

Volume 23 THE AGING BRAIN AND SENILE DEMENTIA
Edited by Kalidas Nandy and Ira Sherwin

Volume 24 CHOLINERGIC MECHANISMS AND PSYCHOPHARMACOLOGY
Edited by Donald J. Jenden

Volume 25 CHOLINERGIC MECHANISMS: Phylogenetic Aspects, Central and Peripheral Synapses, and Clinical Significance
Edited by Giancarlo Pepeu and Herbert Ladinsky

Volume 26 CONDITIONING: Representation of Involved Neural Functions
Edited by Charles D. Woody

Volume 27 THE BASAL GANGLIA: Structure and Function
Edited by John S. McKenzie, Robert E. Kemm, and Lynette N. Wilcock

Volume 28 BRAIN PLASTICITY, LEARNING, AND MEMORY
Edited by B. E. Will, P. Schmitt, and J. C. Dalrymple-Alford

Volume 29 ALZHEIMER'S AND PARKINSON'S DISEASES: Strategies for Research and Development
Edited by Abraham Fisher, Israel Hanin, and Chaim Lachman

Volume 30 DYNAMICS OF CHOLINERGIC FUNCTION
Edited by Israel Hanin

Volume 31 TOBACCO SMOKING AND NICOTINE: A Neurobiological Approach
Edited by William R. Martin, Glen R. Van Loon, Edgar T. Iwamoto, and Layten Davis

Volume 32 THE BASAL GANGLIA II: Structure and Function—Current Concepts
Edited by Malcolm B. Carpenter and A. Jayaraman

Volume 33 LECITHIN: Technological, Biological, and Therapeutic Aspects
Edited by Israel Hanin and G. Brian Ansell

Volume 34 ALTERATIONS IN THE NEURONAL CYTOSKELETON IN ALZHEIMER DISEASE
Edited by George Perry

A Continuation Order Plan is available for this series. A continuation order will bring delivery of each new volume immediately upon publication. Volumes are billed only upon actual shipment. For further information please contact the publisher.

Alterations in the Neuronal Cytoskeleton in Alzheimer Disease

Edited by

GEORGE PERRY
Case Western Reserve University
Cleveland, Ohio

PLENUM PRESS • NEW YORK AND LONDON

Library of Congress Cataloging in Publication Data

Alterations in the neuronal cytoskeleton in Alzheimer disease.
(Advances in behavioral biology; v. 34)

Based on proceedings of a symposium held Dec. 7, 1986 in Washington, D.C.
Includes bibliographies and index.
1. Alzheimer's disease—Pathophysiology—Congresses. 2. Cytoskeletal proteins—Congresses. 3. Cytoplasmic filaments—Congresses. I. Perry, George. II. Series. [DNLM: 1. Alzheimer's Disease—pathology—congresses. 2. Cytoskeletons—pathology—congresses. 3. Neurofibrils—pathology—congresses. WM 220 A466 1986]
RC523.A37 1987 616.8′3 87-29256

ISBN-13: 978-1-4612-8920-3 e-ISBN-13: 978-1-4613-1657-2
DOI: 10.1007/978-1-4613-1657-2

Based on proceedings of a symposium on Reorganization of the Neuronal Cytoskeleton in Aging, held December 7, 1986, in Washington, D.C.

Softcover reprint of the hardcover 1st edition 1987
A Division of Plenum Publishing Corporation
233 Spring Street, New York, N.Y. 10013

TO THE MEMORY OF MY FATHER

PREFACE

The neuronal cytoskeleton is a complex structure responsive to both intrinsic and extrinsic factors. Defined populations of neurons in the brains of patients with Alzheimer and several other neurodegenerative diseases contain abnormal filamentous accumulations which share elements with the cytoskeleton. Although there is a general consensus that these abnormal filaments do contain cytoskeletal elements, much debate remains regarding which cytoskeletal elements are incorporated and whether the cytoskeletal rearrangement is primary or secondary to other cellular changes.

In this book these questions are addressed in a historical perspective in light of new data that allows the reinterpretation of previously reported results. Contributions are based on many of the major techniques of modern biology including biochemistry, molecular biology, electron microscopy and immunocytochemistry.

In the view of the editor, this volume is being written at a time when our understanding of the cytopathology of Alzheimer disease is moving from predominantly descriptive to both analytical and mechanistic. I hope that this contribution will provide impetus to speed this transition.

George Perry

Cleveland, Ohio

ACKNOWLEDGEMENT

The support of the Fidia Pharmaceutic Corporation for the computer-generated color figure on page 65 is gratefully acknowledged.

CONTENTS

NEUROFILAMENTS: A REVIEW AND UPDATE

Steven S.M. Chin and Ronald K.H. Liem

Department of Pharmacology
New York University School of Medicine
New York, N.Y.

INTRODUCTION

For decades neuroanatomists have been familiar with a distinctive fibrillar appearance of large mammalian axons after treatment of nervous tissue with various silver stains. These structures were aptly referred to as "neurofibrils." It was only relatively recently through combined efforts utilizing electron microscopic, biochemical, and immunological methodologies that these well-known structures have been more precisely defined and characterized. Today we refer to these "neurofibrils" as neurofilaments. The goal of this chapter is to review our current knowledge and understanding of neurofilaments in the context of normal cell function.

MORPHOLOGY

Electron microscopic examination of ultra-thin sections taken through large mammalian axons clearly reveal an abundance of electron-dense filaments measuring from 7 to 15 nanometers (nm) in diameter[1]. These neurofilaments (NF) are easily distinguished from the larger (25-30nm) microtubules. Ultra-thin sectioned material also reveal a finer fibrillar, whisker-like material which seems to crosslink neurofilaments to each other as well as to microtubules. This feature is unique among the 10nm or intermediate filaments (IF) of which the neurofilaments are a member. Studies using quick-freeze, deep-etching techniques have confirmed the existence of these crosslinkers (4-6nm in diameter and 20-50nm in length) *in situ*[2,3]. High magnification analysis of cross-sectioned neurofilaments reveal an electron-lucent core measuring approximately 3nm in width. Metal-shadowed preparations of isolated and re-assembled neurofilaments have consistently demonstrated a 21nm axial periodicity which has been directly correlated with the prevailing model of IF structure (see below)[4,5].

BIOCHEMISTRY

Mammalian NF proteins have been isolated and characterized as three distinct polypeptides often referred to as the neurofilament triplet. The apparent molecular masses of the individual triplet polypeptides as determined by SDS-PAGE vary depending on the particular species from which the proteins were isolated but usually cluster around 200-, 150-, and 70-kilodaltons (kD). The molecular masses of the NF proteins vary from species to species (especially when comparing between taxonomic classes or among different phyla) and the larger NF proteins run anomalously on SDS-PAGE[6]. The individual polypeptides are therefore referred to as NF-L (low or light), NF-M (middle or medium), and NF-H (high or heavy).

Structure

Structurally, the neurofilament proteins conform to the general topographical model presently adopted for the IF family of proteins (for reviews see [7,8]). This model consists of a three domain structure containing a highly conserved central alpha-helical domain flanked by variable N-terminal head and C-terminal tail domains. The central alpha-helical domain consists of a 40-kD (~310 amino acid residues) rod region which is composed of two alpha-helical coils each about 21-22nm long, referred to as coils 1 and 2, separated by a non-helical, potentially beta-sheet, linker region (known as segment L12). Within coils 1 and 2 the amino acid residues are arranged into a quasi-repeating heptad motif of the form $(a\text{-}b\text{-}c\text{-}d\text{-}e\text{-}f\text{-}g)_n$, where a and d are frequently apolar residues. In this arrangement a stripe of apolar residues is formed winding down along the alpha-helix. It is probably through the interaction of this apolar stripe in which the basic IF dimeric coiled-coil structure is formed. Coils 1 and 2 are further divided into smaller segments through breaks in the heptad repeat. These segments are known as segments 1A (~5 heptads) and 1B (~14-15 heptads) separated by linker segment L1, and segment 2A (~2-3 heptads) and 2B (~17 heptads) separated by linker segment L2. Additionally, segment 2B contains a conserved reversal of heptad repeats forming a distinct discontinuity or "stutter" located near the center of this segment. Noteworthy are two highly conserved regions within this central rod region which are located in the second and third heptads of segment 1A and in the last 4-5 heptads of segment 2B. In fact, Geisler et al.[9] have shown that the epitope recognized by the monoclonal antibody IFA described by Pruss et al.[10] (which reacts with all IF proteins) resides within the last 20 residues at the end of coil segment 2B.

It is within the head and tail domain where variable sequences reside which account for the individuality of IF polypeptides. Utilizing the central rod domain sequence as a base and taking into consideration deviations in the coil sequences as well as the hypervariable head and tail domains, the IF proteins are subdivided into five types[7,8]. The small acidic and large basic or neutral keratins are known as Type I and Type II IFs, respectively. Vimentin, desmin, and glial fibrillary acid proteins are Type III IFs. Available sequences on mammalian NF sequences have warranted their separation into their own group - Type IV. Finally, the nuclear lamin proteins, which have recently been found by sequence analysis and assembly studies to be IF proteins[11,12,13], have been classified as Type V IFs because of unique features in their rod domains.

The basic IF molecule, as determined by cross-linking studies is a coiled-coil homodimer with the IF monomer subunits oriented parallel and in-register with one another[7,8,14]. The monomer chains are most probably held together by hydrophobic interactions along an apolar stripe formed along the alpha-helical rod domain by the a and d residues of the heptad repeats. These homodimers associate with one another to form the fundamental four-chain coiled-coil unit common to all IFs. There is much more uncertainty and therefore much less agreement concerning the exact nature of the tetramer unit and how it packs into an IF structure[7,8]. However, it is accepted that the core filament structure is composed of a packed array of coiled-coil tetrameric subunits arranged such that the end domains are located on the surface of the filament.[7,15] This arrangement is particularly significant in the case of the neurofilaments since the NF-M and NF-H subunits both possess extensive C-terminal tails[9,17,18]. The packing can be arranged such that these tails are peripherally located and the coiled-coils are located in the backbone core in a regular manner, which also explains the "whispy" surface structures observed on isolated neurofilament preparations. Additionally, this situation would also be consistent with the peripheral staining pattern obtained by Willard and Simon[19] when they decorated neurofilaments with subunit specific neurofilament antibodies.

<u>Assembly</u>

A general property of purified IF proteins is their ability to polymerize into 10nm curvilinear structures under the appropriate buffer conditions. The neurofilament triplet has similarly been shown to reassemble into 10nm filamentous structures[20-22]. However, of the individual subunits only NF-L is capable of forming true intermediate type filaments in the absence of the other two subunits. Under reassembly conditions NF-M and NF-H can form sedimentable material composed of aggregates and short filaments but not intermediate filaments, although Gardner et al.[23] have reported the formation of 10nm filament structures from bovine NF-M. If NF-L is co-assembled with either NF-M or NF-H, normal appearing filaments are formed which upon analysis reveals the presence of NF-L and the corresponding NF polypeptides with which it was co-assembled. Results of such experiments firmly establish that NF-L forms the core structure of NF and that NF-M and NF-H co-assembly is dependent upon the presence of NF-L. It has also been demonstrated that once urea solubilized NF-L is assembled into filaments subsequent incorporation of NF-M or NF-H is not observed[24]. These results combined with the model of IF structure would support the view that NF-M and NF-H are integral components of normal neurofilaments, assembled into the core NF backbone via their central coiled-coil domains. The presence of their unique long tail domains could explain their inability to assemble into pure NF-M or NF-H IF structures. The presence of the NF-L polypeptide may be necessary to neutralize the assembly-negating effects of these tail domains. The estimated average molar ratio of NF-H:NF-M:NF-L of 1:2.5:9 for intact neurofilaments[24] is supportive of this notion if uniform distribution of NF subunits within a NF is assumed since the more numerous NF-L molecules would tend to space the NF-M and NF-H subunits further apart from one another. Additionally, Kaufmann et al.[25] have demonstrated that amino-terminal sequences in the desmin molecule are necessary for normal IF forming ability while a desmin fragment lacking the carboxy-terminal 27 residues is assembly-competent. It could also be argued that the head domains of NF-M and NF-H are similarly responsible for the observed assembly-incompetence, although sequence data demonstrate a fair degree of homology between the head domains of NF-L and NF-M[17].

Recently it has been shown that the phosphorylation state of the NF subunits plays little if any role in assembly[26]. In fact, Black et al.[27] have demonstrated that <u>in vivo</u> phosphorylation of the NF subunits takes place after newly synthesized NF protein is incorporated into the insoluble cytoskeleton.

<u>Interactions</u>

Study of the interactions of NF with the cellular milieu have provided insights into possible functions of NFs. From early morphologic studies it was apparent that NFs interact with one another and with other axoplasmic structures (especially microtubules) via thin filamentous structures. These cross-bridging structures have been assumed to be composed of tail domains of neurofilaments M and H and of microtubule-associated proteins[2,19,28-31]. Immunoelectron microscopic staining and freeze-etch studies have clearly demonstrated the peripheral nature of the NF-H tail domain. MAP 2 has also been demonstrated to associate with neurofilament and tubulin proteins in β,β'-iminodiproprionitrile (IDPN)-treated axons in which neurofilaments and microtubules are induced to become spatially separated within the axoplasmic compartment[32]. Biochemically it has been shown that neurofilaments in the presence of MAPs can form stable associations with microtubules[33]. More detailed binding studies have resulted in a surprising finding. MAP 2 and tau proteins bind directly to the core NF-L subunit, not to the supposedly peripherally located NF-M and NF-H subunits[29]. These findings suggest that some of the observed cross-bridges are MAPs directly associating with neurofilament cores and interacting with neighboring microtubules.

<u>Phosphorylation</u>

Neurofilament polypeptides like other intermediate filament proteins undergo phosphorylation both <u>in vivo</u> and <u>in vitro</u>[34]. All three polypeptides of mammalian NF are phosphorylated although with varying amounts of phosphate. Studies of rat and bovine neurofilament have resulted in the following phosphate values Table One:

Table 1: Number of phosphate groups/NF subunit as determined by different laboratories[35-38].

NF Subunit:	bovine[35]	bovine[36]	bovine[37]	rat[38]
	mol phosphate/ mol NF subunit			
NF-H	104	25	19	22
NF-M	26	12	7	9
NF-L	<1	3	1	3

The values for NF-H and NF-M are based on the molecular weights determined by SDS-PAGE, which are overestimated by at least 50%. The actual moles of phosphate/mol of NF-H or NF-M subunits are therefore about 2/3 of the reported values.

Invertebrate neurofilament polypeptides from loligo and myxicola also are phosphorylated to varying extents demonstrating the generality of neurofilament phosphorylation[39].

Phosphopeptide analysis demonstrates that NF is predominantly phosphorylated on serine residues although phosphothreonine has been detected in both bovine and rat NF[37,38]. These phosphate groups are located mainly in the tail domains of NF-H and NF-M and many of these groups are accessible to enzymatic cleavage by acid and alkaline phosphatase. However, phosphates in NF-L are generally inaccessible to phosphatase treatment possibly because NF-L's smaller tail domain may be conformationally packed closer to the core filament structure or because NF-L is phosphorylated in the rod region. NF-M and NF-H also contain residual phosphate groups which are resistant to phosphatase treatment. Upon progressive dephosphorylation, NF-M and NF-H demonstrate a dramatic successive increase in mobility on SDS-PAGE[37,38]. This mobility shift provides evidence that phosphorylation directly influences the conformation of the NF-M and NF-H molecules. Recent data suggest that the anomalous behavior on SDS-PAGE is due to the combined effects of phosphorylation state and to stretches of glutamic acid rich regions of the tail domain which could confer extended helical properties to the region[40].

Antibodies specific for phosphorylated and dephosphorylated NF have been used to confirm this phosphorylation-dependent shift and have also demonstrated that unphosphorylated NF isoforms exist in vivo. Using these antibodies, Sternberger and Sternberger[41] have immunohistochemically demonstrated the presence and differential localization of both phosphorylated and unphos-phorylated NF forms in brain tissue. Unphosphorylated NF is preferentially located in neuronal perikarya while phosphorylated forms are found predominantly in axonal processes. Bennett and DiLullo[42] and Black et al.[27] have demonstrated that cultured neurons incorporate unphosphorylated NF into the cytoskeleton and it is only after these new NF polypeptides are assembled onto the existing IF cytoskeleton are they phosphorylated. A similar result has been shown for vimentin[43]. Furthermore, Georges et al.[26] have demonstrated that in vitro dephosphorylated porcine neurofilament subunits are capable of reassembling into normal IF structures. Therefore, the phosphorylation of NF subunits does not seem to play a role in neurofilament assembly. This is consistent with the finding that the reassembly properties of keratins, desmin, and vimentin are not related to their phosphate content[44].

Speculation as to the function of neurofilament phosphorylation has mainly been focused on the direct correlation of heavy phosphate content of NF-H with its proposed cross-bridging function. The possibility exists that gradual post-translational phosphorylation of nascent NF-M and NF-H molecules as they are transported from perikarya into the proximal axonal region and further into the distal axon may play a role in the formation and regulation of the extended cross-linker structures[45].

In vitro studies have demonstrated both cAMP-dependent and -independent protein kinases which may play a role in the in vivo phosphorylation of neurofilament polypeptides. Leterrier et al.[46] demonstrated that a cAMP-dependent microtubule-associated protein kinase preferentially phosphorylates bovine and rabbit NF-M in vitro. Julien et al.[47] have described a cAMP-independent protein kinase activity associated with isolated rat neurofilaments. It was shown that NF-M was again the preferred substrate although phosphopeptide mapping indicated that the in vitro phosphorylated NF-H contained phosphothreonine which is normally not found in isolated NFs. The physiological significance of both these protein kinase activities still remains to be elucidated.

Degradation

Protein degradation and turnover is a vital cellular function which enables normal cells to constantly replenish cytoplasmic pools of important proteins. Neurofilament protein are continuously being synthesized in neuronal cell bodies and transported into axonal processes[27]. Presumably these neurofilament proteins are degraded either as they migrate down axons or at their distal terminals since accumulation of these proteins is not normally observed in synapses.[1] Recycling of these proteins is unlikely since neurofilament proteins have not been reported to be retrogradely transported. The exact mechanisms of neurofilament protein turnover are still unknown. However, Ca^{2+}-induced disintegration of neurofilament structures has been well documented[48,49] and has motivated researchers to concentrate studies on calcium-activated neutral proteinases (CANPs). CANPs isolated from loligo axoplasm and mammalian brain tissue demonstrate distinct activities directed toward the modification and degradation of neurofilament proteins[50-57]. Additionally CANPs have also been demonstrated to degrade other intermediate filament proteins such as vimentin, desmin, and glial fibrillary acidic protein[52-58] suggesting a common mechanism in which non-keratin intermediate filaments are degraded.

Calcium-activated neutral proteins (EC 2.4.22.17; also known as calpains) are a family of cysteine endopeptidases found in a wide variety of vertebrate and invertebrate tissues[59-61]. Two types of CANPs have been characterized. One type is most active under conditions of micromolar concentrations of Ca^{2+} and is referred to as μCANP, CANP I,

CANP B, or calpain I. The other type, referred to as mCANP (or CANP II, CANP A, or calpain II) is active at millimolar concentrations of Ca^{2+}.

At present it appears that the observed Ca^{2+}-induced disintegration of neurofilament structures could be explained by the presence of neuronal CANPs. It has been hypothesized that as neurofilament proteins are transported into the distal axon and reach the axonal terminals, local Ca^{2+} concentration changes would activate CANPs located in the synaptic terminals and thereby degrade the proteins[62]. This would be consistent with the reported disappearance of radiolabelled neurofilament protein after their arrival into the terminals[63]. Leupeptin which inhibits CANP activity has been shown to induce neurofilament accumulation in synaptic terminals if it is injected into neurons[64]. However, the mechanism by which mCANP functions is not known since endogenous Ca^{2+} concentrations are much lower than its optimal value for maximal activity as measured in vitro. It can be argued that μCANP is active at endogenous Ca^{2+} concentrations but, at least in mice, it has been well demonstrated that it specifically post-translationally modifies the NF-M subunit, failing to degrade the NF-M subunit further or to degrade the other NF subunits[56,65]. Further, a recent immunohistochemical study of the localization of μCANP and mCANP in rat brain failed to demonstrate the presence of either enzyme in synaptic terminals[66]. Immunoreactive staining for μCANP was mainly observed in the neuronal cell bodies while mCANP was mainly localized in glial cell bodies.

Clearly, these discrepancies must be resolved and much more research needs to be done before neurofilament protein turnover can be fully understood. Presently we can only conclude that neuronal CANPs potentially play a role in normal neurofilament protein degradation but we cannot exclude the probable participation of other as yet unidentified enzymes and protein degradation systems. For example, human brain cathepsin D, a lysosomal enzyme, has been shown to degrade neurofilament proteins although the physiological significance of this finding is still unclear[67].

MOLECULAR BIOLOGY

Amino acid sequence analysis has been instrumental in elucidating the primary structure of the IF proteins[8]. This approach however is tedious and difficult to complete on large molecules such as the NF polypeptides. Molecular cloning has greatly simplified the task of deducing the primary amino acid sequences of many protein molecules including those of the IF molecules. Considerable sequence data is now available on the Type I and II keratins and the Type III IFs[68-74]. However until very recently there was limited amino acid or nucleic acid sequence data for the NF proteins -- most if not all of the available amino acid sequence data on NF proteins were the result of the excellent work done by the group of Geisler et al.[9,17,75].

In 1985 Lewis and Cowan[76] reported the cloning of a cDNA probe encoding the mouse NF-L. Later that year Julien et al.[77] reported on a rat NF-L cDNA. These two reports confirmed the amino acid sequence of porcine NF-L as reported by Geisler et al.[75] and demonstrated that the amino acid sequence of porcine, mouse, and rat NF-L protein are highly conserved (with >93% identity). We recently have cloned and sequenced a cDNA clone that fully encodes the rat NF-L protein (Chin, Wu and Liem, unpublished data). We have also recently cloned and fully sequenced two rat NF-M cDNA clones, one of which encodes the entire rat NF-M polypeptide[40]. The cloning of cDNAs for NF-H have only appeared as preliminary reports[78,79].

Analysis of the available amino acid and nucleic acid sequence data have resulted in the following findings:

1) rodent NF amino acid sequences are highly homologous to one another as well as to those of porcine NF;

2) the amino acid sequences conform to the three domain coiled-coil model of IF structure;

3) tail domains of NF-L, NF-M, and NF-H consistently demonstrate stretches of glutamic acid-rich regions;

4) the predicted molecular masses of unphosphorylated rat NF-M is 95.4-kD, which is considerably smaller than the molecular mass obtained by SDS-PAGE (145-kD for rat NF-M); and

5) the peptide sequence of Lys-Ser-Pro is found many times in the tail domain of NF-M and NF-H and could possibly be the phosphorylation sites.

Lewis and Cowan[80] have extended their studies of NF-L into those of NF-L gene structure. On comparing the mouse NF-L gene structure with those of Type I and Type II keratins, desmin, vimentin, and glial fibrillary acidic protein, Lewis and Cowan[80] found that the number and placement of the introns were completely anomalous. Whereas the other members of the IF family demonstrate an homologous positioning of seven or more introns within their gene structure, NF-L demonstrates the presence of only 3 introns in positions corresponding to none of the above characterized genes. On the other hand, the NF-M gene structure shows the presence of two introns[81], corresponding to the first of the two introns in the NF-L gene. These studies confirm the classification of the neurofilaments as a separate class of intermediate filaments.

CELL BIOLOGY

Differential localization and expression

Immunohistochemical studies of various neural tissues and cells have demonstrated neurofilament expression to be a complex event. Neurons from different parts of the nervous system may express varying amounts of neurofilament protein. Large neurons consistently demonstrate neurofilament immunoreactivity while certain smaller neurons, especially those in the CNS, display little or no detectable NF proteins.[82] Specifically, dorsal root ganglion (DRG) neurons illustrate this characteristic. DRGs are composed of two types of sensory neurons, large light or A cells and small dark or B cells. Immunohistochemical studies have clearly demonstrated rich neurofilament triplet protein staining in the large light neurons while the small dark neurons show little if any staining[83-85].

Generally, NF staining of axonal processes is more intense than that for the neuronal cell soma. This phenomena is due to: 1) most polyclonal and monoclonal NF antibodies are predominantly directed against phosphorylated epitopes present primarily in the axons; and 2) axons contain much more neurofilaments than the cell soma[28,86]. Antibodies specifically directed against unphosphorylated NF epitopes have demonstrated NF immunoreactivity only in the perikaryal cytoplasm[41]. Previously observed discrepancies in the apparent uniform or non-uniform distribution of the neurofilament triplet within neurons could be explained by the use by various research groups of different NF antibodies directed toward the same NF subunit but specific for a particular phosphorylation site. The initial report of a uniform distribution of the three NF proteins could have been due to the use of antibodies that recognized both phosphorylated and unphosphorylated epitopes[87]. Subsequent reports of specific localization or absence of the NF-M or NF-H subunits in either the neuronal soma or axon[28,88,89] could be explained by the use of antibodies that predominantly recognized either the phosphorylated or unphosphorylated forms of the NF-M or NF-H proteins[90].

Developmental studies of avian and mammalian nervous systems have demonstrated a differential expression of intermediate filaments in embryonic neurons[91-97]. In general, as neuroblasts differentiate and mature they appear to sequentially express vimentin, then NF-L and NF-M, and finally NF-L, NF-M and NF-H together. Transient co-existence of vimentin and the NF proteins has been well documented for many embryonic neural cell systems[93,95,96]. Curiously, the axonless horizontal cells of adult mouse and rat retina continue to express vimentin together with the NF triplet[95,98,99]. The occurrence of NF-L and NF-M is consistently observed to be expressed developmentally earlier than that of NF-H, which appears only after the neurons have matured[28]. Axons of the peripheral nervous system demonstrate NF-H immunoreactivity much earlier than those of the more slowly developing central nervous system. Additionally, the developmentally precocious chick possesses NF-H containing axons much earlier in pre- and post-natal periods than similarly aged mammalian species (Chin and Liem, unpublished observations).

Neurofilament expression is normally restricted to neurons and cells of the diffuse neuroendocrine system[82,100]. The report by Granger and Lazarides[101] of a NF-L-like protein being expressed by adult avian erythrocytes is the only exception to this generalization. Interestingly, Bennett and DiLullo[102] have observed transient expression of NF-M protein in early neural and nonneural (e.g., cardiac myoblasts) cells of embryonic chicks. The significance of such a finding is unclear. Neurofilament proteins have been reported to be expressed in various tumors[103,104] and tumor derived cell lines. For example, 1) certain mouse neuroblastoma derived cell lines express the NF-L subunit[105,106], 2) rat PC12 cells (a cell line derived form a pheochromocytoma) can be induced with nerve growth factor to express the neurofilament triplet[107,108], 3) the clonally derived, pluripotent human embryonal carcinoma cell line NTERA-2cl.D1 can be induced with retinoic acid to differentiate into cells that express the neurofilament triplet[109], and 4) variant small cell lung cancer cell lines NC1-H82 and NC1-H524 demonstrate co-expression of NF-L, NF-M and vimentin[110]. The use of these cell lines in conjunction with primary neuronal cultures will be useful in future studies of neurofilament gene expression.

Phylogenetic expression

Neurofilament expression has been studied in a variety of vertebrate and invertebrate species[86,111-113]. Mammalian and avian species have been most extensively studied of the vertebrates and include species such as rat, mouse, hamster, gerbil, guinea pig, rabbit, pig, cow, ox, cat, baboon, man, chicken, parakeet, finch, canary, and goose. Of the invertebrates neurofilaments from the marine fan-worm, myxicola, and the squid, loligo, have been thoroughly characterized. Neurofilaments from both of these invertebrates consist of two distinct polypeptides -- myxicola NF appears as 160-kD and 150-kD polypeptides and loligo NF appears as 200-kD and 60-kD polypeptides. The molluscan Aplysia has also been shown to possess a neurofilament doublet of 65-kD and 60-kD. This is unlike the situation seen with mammalian and avian neurofilaments which consistently appear as a triplet. Subsequent studies have been performed to establish and characterize the expression of neurofilament proteins in lower vertebrates and invertebrates. During the course of these studies certain criteria were found to be useful in establishing the authenticity of certain polypeptides as neurofilament proteins. Basic criteria included the following: 1) demonstration of immunoreactivity with the general IFA monoclonal antibody that has been shown to react with all IF proteins[10]; 2) demonstration of specific expression of the proteins in neurons and its transport into axons; and 3) demonstration of staining with the Bodian silver staining technique which has been shown to be specific for neurofilament proteins[113-115]. Using these criteria, putative NF proteins have been identified in a large number of organisms. Among the invertebrates, examples from phyla annelida and mollusca have been shown to possess neurofilaments. Species from phylum arthropoda, however, have consistently demonstrated an absence of IF proteins in nerve axons[113]. The absence of cross-hybridizing genomic DNA sequences in Drosophila probed with a mouse NF-L cDNA clone[76] support the view that arthropods do not express NF proteins.

Lower vertebrates have been shown to express neurofilament protein. In general, reptilian NF proteins occur as a doublet while amphibian NF proteins appear both as triplets and doublets, depending on the particular species. Lasek et al.[113] have studied neurofilament expression in a number of bony and cartilaginous fish and have found that their NF proteins vary from doublets to quadruplets, again dependent on the particular species of fish studied. Further studies of species-dependent expression of neurofilament proteins and their molecular structure will be necessary before the evolution of these proteins can be fully elucidated.

Axonal transport

Axonal transport studies have substantially contributed to the understanding of the dynamics of neurofilaments. The neurofilament triplet proteins are co-transported together as a major component of the slowest component of axonal transport also known as slow component a (SCa) or group V (for detailed reviews of axonal transport see[116,117]). SCa components typically move at a velocity of 0.1-1 mm/day while the more numerous slow component b (SCb) proteins move ahead at a velocity of 1-5 mm/day. It appears that both components of slow axonal transport are responsible for the continuous anterograde flow of axoplasmic matrix through neuronal axons. The cohesive nature of SCa and SCb transport has led to the hypothesis that the individual constituents of the axoplasmic matrix are translocated through the axon as a macromolecular complex[117,118]. For example, the neurofilament triplet is co-transported with the tubulins as the major constituents of SCa. This further strengthens the view that neurofilaments and microtubules associate intimately with each other to form a basis for the neuronal axoplasm.

Many of the individual constituents of SCa and SCb have been identified[117,119] and their transport characteristics have been compared in different neurons[120,121]. The results of such comparisons have established the existence of cytotypic differences in both composition and transport velocity of the axonally transported cytoskeleton and cytomatrix[120,121]. For example, tubulin was found to be the major SCb protein in ventral motor neuron (VMN) and dorsal root ganglion (DRG) axons, but was not found to be present in SCb in optic nerve (ON) axons. The transport velocities for SCa and SCb proteins were also substantially faster in the peripheral VMN and DRG axons than in the central ON axons. Despite these and other reported variances the neurofilament triplet is consistently transported in the slowest transport component in both central and peripheral neurons. The only significant difference involving neurofilaments was a noticeably higher ratio of neurofilament protein to tubulin in SCa in DRG and VMN axons compared to ON axons[121]. These and other observed cytotypic differences subtlely reflect the diversity of form and function exhibited by central and peripheral neurons.

The cytoskeletal components of axons such as actin, tubulin, and neurofilaments have also been hypothesized to play a role in axonal transport. The broad actin wavefront seen in transport studies has been interpreted to possibly be due to its interaction with the components of the subaxolemmal matrix.[120,121] SC velocities could possibly be regulated through such interactions with those involving cytomatrix components. Microtubules have been shown to be involved in fast axonal transport[122,123]. Studies using IDPN treatment have clearly demonstrated that vesicular and organellar fast axonal transport is confined to the centrally located microtubular structures and is not associated with the peripherally displaced neurofilament arrays[124-126]. Developmental studies have correlated a decrease in neurofilament transport velocity with the increase in NF-H expression in late post-natal life[127,128]. This find further supports the view that NF-H possesses cross-linking functions and that such cross-linking to other neurofilaments, microtubules, and other undefined structures could modulate the velocity of macromolecular complexes in SCa.

More detailed studies of central neuronal transport have resulted in findings indicative of regional specialization within axons. It was found that the mouse and chick neurofilament middle subunit undergoes post-translational modification as it moves along the length of the axons[45,89]. The 145-kD and 140-kD mouse NF-M protein are observed to be uniformly distributed along the whole of the axon while the 143-kD subunit appears only at the distal end[45]. Nixon and his colleagues[45,129] have also presented evidence that support their hypothesis that neurofilaments exist as two distinct pools. One pool is the normally observed neurofilament component of SCa undergoing continuous transport. The other pool is composed of a relatively stationary neurofilament network. Nixon and Logvinenko[129] estimate that, in the steady state, most of the neurofilament proteins in mouse retinal ganglion cell axons may exist as part of this stationary pool. Furthermore, they report that there exists an increasing proximal to distal gradient of neurofilament protein in mouse retinal ganglion axons. The existence of such a nonuniform distribution of neurofilament protein in axonal processes could further modulate the process of axonal transport through presently undefined interactions.

Morphological studies suggested that one function of neurofilaments is to determine axonal caliber particularly of myelinated axons[130,131]. Experimental manipulations such as axotomy, nerve constriction and crush result in an initial decrease in axonal transport of neurofilament proteins coinciding with a reduction in axonal caliber. As the nerves regenerate neurofilament transport return to normal levels with restoration of axonal caliber[131-133]. Data from experimental galactose neuropathy studies demonstrate that whereas the density of NF in axons is not significantly different from those of the control groups, the decreased number of NFs per axon was correlated with a decrease in axonal caliber[134]. Similar findings were reported for patients with hereditary motor and sensory neuropathy, type I[135]. The consistent correlation between changes in axon caliber and coincident changes in NF number support the view that neurofilament transport plays a role in regulating axonal volume and consequently axonal conduction properties.

PATHOLOGY

A number of degenerative neuropathies are characterized by morphological reorganization of cytoskeletal elements.

Alzheimer's disease[136], idiopathic and post-encephalitic Parkinson's disease[137], amyotrophic lateral sclerosis (ALS)[138,139], Guamanian amyotrophic lateral sclerosis-Parkinsonism-dementia complex[140], Down's syndrome[141], Pick's disease[142], and giant axon neuropathies[143] (GANs) all exhibit a profound alteration of the neuronal cytoskeleton in the form of accumulations of filamentous material within affected neurons. The cytoplasmic accumulation of filaments and/or inclusion bodies in neuropathies associated with these disease states have all been shown to possess neurofilament-immunoreactive antigens[137,141,142,145-147]. Neurofibrillary tangles are found in small numbers in the brains of normally aged men and women but are found in much greater numbers in patients affected with diseases such as Alzheimer's disease, post-encephalitic Parkinson's disease, Down's syndrome, Guamanian amyotrophic lateral sclerosis-Parkinsonism-dementia complex, progressive supranuclear palsy, and subacute sclerosing panencephalitis[146]. These tangles were originally thought to be composed of neurofilaments but are now known to consist of two 10nm filaments wound helically around one another and are referred to as paired helical filaments (PHFs)[148]. The extraordinarily insoluble nature of these fibrillary elements has made it difficult to fully analyze their biochemical composition[149]. However, it has been shown that PHFs contain epitopes found in neurofilaments[141,147,150], microtubule-associated protein 2 (MAP-2)[150,151], and phosphorylated tau proteins[152,153]. Pick bodies[142] and Lewy bodies[137] characteristic of Pick's disease and Parkinson's disease, respectively, are cytoplasmic inclusions composed of filamentous structures which have also been identified as neurofilamentous in nature.

Experimentally-induced neuropathies have been very useful in providing animal model systems with which to study the pathogenesis of the above mentioned diseases[154]. Toxins such as acrylamide,[155] IDPN,[124-126,154] 2,5-hexanedione,[154,156] and aluminum salts[157] seem to affect axonal transport manifesting in accumulations of specific transport components. IDPN appears to inhibit the slow transport of NF protein, producing proximal accumulation of neurofilamentous structures[154,158]. 2,5-Hexanedione, however, appears to accelerate the rate of neurofilament protein transport producing distal accumulations of neurofilaments[156]. Although NF transport appears to contribute to the pathologies, the exact mechanisms responsible for the observed neurofilament accumulations are not known[159]. Factors such as rates of NF synthesis, degradation and post-translational modification (possibly abnormal cross-linking[160,161]) must be taken into consideration. Such experimental systems may be useful as models for studying the proximal and distal types of filamentous accumulations in ALS and GANs[156], respectively.

Autoantibodies reacting with the neurofilament triplet have been identified in kuru and Creutzfeldt-Jakob disease[162-164]. This occurrence has been interpreted as a host reaction to the lysis of neurofilament-loaded neurons. The contribution of such autoantibodies to the pathogenesis of these and other neurodegenerative diseases still remains to be elucidated.

CONCLUDING REMARK

Continued research on neurofilament structure and function will prove to be invaluble to our ever-growing understanding of the workings of the neuron and its relationship with the nervous system as a whole.

ACKNOWLEDGEMENTS

We gratefully thank Drs. David Ennulat, John Peralo, and Emma DiIorio for critically reading the manuscript and their helpful suggestions.

REFERENCES

1. Peters A, de Webster H, Palay S: The Fine Structure of the Nervous System: The Neurons and Supporting Cells. Philadelphia, WB Saunders Co, 1982
2. Hirokawa N: Cross-linker system between neurofilament, microtubules, and membrane organelles in frog axons revealed by the quick-freeze, deep-etching method. J Cell Biol 94:129-142, 1982
3. Schnapp BJ, Reese TS: Cytoplasmic structures in rapid-frozen axons. J Cell Biol 94:667-679, 1982
4. Henderson D, Geisler N, Weber K: A periodic ultrastructure in intermediate filaments. J Mol Biol 155:173-176, 1982
5. Milam L, Erickson HP: Visualization of a 21-nm axial periodicity in shadowed keratin filaments and neurofilaments. J Cell Biol 94:592-596, 1982
6. Kaufmann E, Geisler N, Weber K: SDS-PAGE strongly overestimates the molecular masses of the neurofilament proteins, FEBS Letters 170:81-84, 1984
7. Steinert PM, Parry, DAD: Intermediate filaments: Conformity and diversity of expression and structure. Annu Rev Cell Biol 1:41-65, 1985
8. Weber K, Geisler N: Intermediate filaments: Structural conservation and divergence, Ann NY Acad Sci 455:126-143, 1985
9. Geisler N, Kaufmann E, Fischer S, et al: Neurofilament architecture combines structural principles of intermediate filaments with carboxy-terminal extensions increasing in size between triplet proteins. EMBO J 2:1295-1302, 1983
10. Pruss RM, Mirsky R, Raff MC et al: All classes of intermediate filaments share a common antigenic determinant defined by a monoclonal antibody. Cell 27:419-428, 1981
11. McKeon FD, Kirschner MW, Caput D: Homologies in both primary and secondary structures between nuclear envelope and intermediate filament proteins. Nature 319:463-468, 1986
12. Fisher DZ, Chaudhary N, Blobel G: cDNA sequencing of nuclear lamins A and C reveals primary and secondary structural homology to intermediate filament proteins. Proc Natl Acad Sci USA 83:6450-6454, 1986
13. Gerace L: Nuclear lamina and organization of nuclear architecture, TIBS, November 1986, pp 443-446

14. Parry DAD, Steven AC, Steinert PM: The coiled-coil molecule of intermediate filaments consists of two parallel chains in exact axial register. Biochem Biophys Res Comm 127:1012-1018, 1985
15. Weber K, Geisler N: Intermediate filaments - from wool α-keratins to neurofilaments: A structural overview, in Cancer Cells 1, The Transformed Phenotype, Cold Spring Harbor, Cold Spring Harbor Lab, 1984, pp 153-159
16. Geisler N, Kaufmann E, Weber K: Antiparallel orientation of the two double-stranded coiled-coils in the tetrameric protofilament unit of intermediate filaments. J Mol Biol 182:173-177, 1985
17. Geisler N, Fischer S, Vandekerckhove J, et al: Hybrid character of a large neurofilament protein (NF-M): Intermediate filament type sequence followed by a long and acidic carboxy-terminal extension. EMBO J 3:2701-2706, 1984
18. Geisler N, Fischer S, Vandekerckhove J, et al: Protein-chemical characterization of NF-H, the largest mammalian neurofilament component; intermediate filament-type sequences followed by a unique carboxy-terminal extension. EMBO J 4:57-63, 1985
19. Willard M, Simon C: Antibody decoration of neurofilaments. J Cell Biol 89:198-205, 1981
20. Geisler N, Weber K: Self-assembly *in vitro* of the 68,000 molecular weight component of the mammalian neurofilament triplet protein into intermediate-sized filaments. J Mol Biol 151:565-571, 1981
21. Liem RKH, Hutchinson SB: Purification of individual components of the neurofilament triplet: Filament assembly from the 70,000-dalton subunit. Biochemistry 21:3221-3226, 1982
22. Zackroff RV, Idler WW, Steinert PM, et al: *In vitro* reconstitution of intermediate filament from mammalian neurofilament triplet polypeptides. Proc Natl Acad Sci USA 79:754-757, 1982
23. Gardner EE, Dahl D, Bignami A: Formation of 10-nanometer filaments from the 150 k-dalton neurofilament protein *in vitro*. J Neurosci Res 11:145-155, 1984
24. Minami Y, Endo S, Sakai H: Participation of 200K or 150K subunit of neurofilament in construction of the filament core with 70K subunit and promotion of tubulin polymerization by incorporated 200K subunit. J Biochem 96:1481-1490, 1984
25. Kaufmann E, Weber K, Geisler N: Intermediate filament forming ability of desmin derivatives lacking either the amino-terminal 67 or the carboxy-terminal 27 residues. J Mol Biol 185:733-742, 1985
26. Georges E, Lefebvre S, Mushynski WE: Dephosphorylation of neurofilaments by exogenous phosphatases has no effect on reassembly of subunits. J Neurochem 47:477-483, 1986
27. Black MM, Keyser P, Sobel E: Interval between the synthesis and assembly of cytoskeletal proteins in cultured neuron. J Neurosci 6:1004-1012, 1986
28. Hirokawa N, Glicksman MA, and Willard MB: Organization of mammalian neurofilament polypeptides within the neuronal cytoskeleton. J Cell Biol 98:1523-1536, 1984

29. Heimann R, Shelanski ML, Liem RKH: Microtubule-associated protein bind specifically to the 70-kDa neurofilament protein. J Biol Chem 260:12160-12166, 1985
30. Miyata Y, Hoshi M, Nishida E, et al: Binding of microtubule-associated protein 2 and tau to the intermediate filament reassembled from neurofilament 70-kDa subunit protein. J Biol Chem 261:13026-13030, 1986
31. Bloom GS, Luca FC, Vallee RB: Cross-linking of intermediate filaments to microtubules by microtubule-associated protein 2. Ann NY Acad Sci 455:18-31, 1985
32. Papasozomenos SC, Binder LI, Bender PK et al: Microtubule-associated protein 2 within axons of spinal motor neurons: associations with microtublues and neurofilaments in normal and β,β'-iminodipropionitrile-treated axons. J Cell Biol 100:74-85, 1985
33. Aamodt EJ, Williams RC: Microtublue-associated protein connect microtubules and neurofilaments in vitro. Biochemistry 23:6023-6031, 1984
34. Lazarides E: Intermediate filaments: A chemically heterogenous, developmentally regulated class of proteins. Annu Rev Biochem 51:219-250, 1982
35. Jones SM, Williams RC: Phosphate content of mammalian neurofilaments. J Biol Chem 257:9902-9905, 1982
36. Wong J, Hutchison SB, Liem RKH: An isoelectric variant of the 150,000-dalton neurofilament polypeptide: Evidence that phosphorylation state affects its association with the filament. J Biol Chem 259:10867-10874, 1984
37. Carden MJ, Schlaepfer WW, Lee VMY: The structure, biochemical properties, and immunogenicity of neurofilament peripheral regions are determined by phosphorylation state. J Biol Chem 260:9805-9817, 1985
38. Julien JP, Mushynski WE: Multiple phosphorylation sites in mammalian neurofilament polypeptides. J Biol Chem 257:10467-10470, 1982
39. Shecket G, Lasek RJ: Neurofilament protein phosphorylation: Species generality and reaction characteristics. J Biol Chem 257:4788-4795, 1982
40. Napolitano EW, Chin SSM, Colman DR, et al: Characterization, in vitro expression and complete amino acid sequence of rat NF-M, the middle molecular weight neurofilament protein, using specific cDNA clones. J Neurosci in press
41. Sternberger LA, Sternberger NH: Monoclonal antibodies distinguish phosphorylated and nonphosphorylated forms of neurofilaments in situ. Proc Natl Acad Sci USA 80:6126-6130, 1983
42. Bennett GS, DiLullo C: Slow posttranslational modification of a neurofilament protein J Cell Biol 100:1799-1804, 1985
43. Blikstad I, Lazarides E: Vimentin filaments are assembled from a soluble precursor in avian erythroid cells. J Cell Biol 96:1803-1808,1983
44. Steinert PM, Wantz ML, Idler WW: O-Phosphoserine content of intermediate filament subunits. Biochemistry 21:177-183, 1982

45. Nixon RA, Brown BA, Marotta CA: Posttranslational modification of a neurofilament protein during axoplasmic transport: Implications for regional specialization of CNS axons. J Cell Biol 94:150-158, 1982
46. Leterrier JF, Liem RKH, Shelanski ML: Preferential phosphorylation of the 150,000 molecular weight component of neurofilament by a cyclic AMP-dependent, microtubule-associated protein kinase. J Cell Biol 90:755-760, 1981
47. Julien JP, Smoluk GD, Mushynski WE: Characterization of the protein kinase activity associated with rat neurofilament preparations. Biochem Biophys Acta 755:25-31, 1983
48. Schlaepfer WW: Calcium-induced degeneration of axoplasm in isolated segment of rat peripheral nerve. Brain Res 69:203-215, 1974
49. Schlaepfer WW: Structural alterations of peripheral nerve induced by the calcium ionophore A23187. Brain Res 136:1-9, 1977
50. Pant HC, Terakawa S, Gainer H: A calcium activated protease in squid axoplasm. J Neurochem 32:99-102, 1979
51. Pant HC, Gainer H: Properties of a calciun-activated protease in squid axoplasm which selectively degrades neurofilament proteins. J Neurobiol 11:1-12,1980
52. Schlaepfer WW, Zimmerman VJP: Calcium-mediated breakdown of glial filaments and neurofilaments in rat optic nerve and spinal cord. Neurochem Res 6:243-255, 1981
53. Zimmerman VJP, Schlaepfer WW: Characterization of a brain calcium-activated protease that degrades neurofilament proteins. Biochemistry 21:3977-3983, 1982
54. Ishizaki Y, Tashiro T, Kurokawa M: A calcium-activated protease which preferentially degrades the 160-kDa component of the neurofilament triplet. Eur J Biochem 131:41-45, 1983
55. Gallant PE, Pant HC, Pruss RM, et al: Calcium-activated proteolysis of neurofilament proteins in the squid giant neuron. J Neurochem 46:1573-1581, 1986
56. Nixon RA, Quackenbush R, Vitto A: Multiple calcium-activated neutral proteinases (CANP) in mouse retinal ganglion cell neurons: Specificities for endogenous neuronal substrates and comparison to purified brain CANP. J Neurosci 6:1252-1263, 1986
57. Vitto A, Nixon RA: Calcium-activated neutral proteinases of human brain: Subunit structure and enzymatic properties of multiple molecular forms. J Neurochem 47:1039-1051, 1986
58. Nelson WJ, Traub P: Purification and further characterization of the Ca^{2+}-activated proteinase specific for the intermediate filament proteins vimentin and desmin. J Biol Chem 257:5544-5553, 1982
59. Ishiura S: Calcium-dependent proteolysis in living cells. Life Sci 29:1079-1087, 1981
60. Kishimoto A, Kajikawa N, Tabuchi H, et al: Calcium-dependent neutral proteases, widespread occurrence of a species of protease active at lower concentrations of calcium. J Biochem 90:889-892, 1981

61. Murachi T: Intracellular Ca^{2+}-protease and its inhibitor protein, calpain and calpastatin, in Cheung WY (ed): Calcium and Cell Function, Vol 4, New York, Academic Press, 1983, pp 377-419
62. Lasek RJ, Black MM: How do axons stop growing? Some clues from the metabolism of the protein in the slow component of axonal transport, in Roberts S, Lajtha A, Gispen WH (eds): Mechanism, Regulation and Special Functions of Protein Synthesis in the Brain, New York, Elsevier-North Holland Biomedical Press, 1977, pp 161-169
63. Lasek RJ, Hoffman PN: The neuronal cytoskeleton, axonal transport and axonal growth, in Goldman R, Pollard T, Rosenbaum E (eds): Cell Motility, New York, Cold Spring Harbor Laboratory 1976 pp 1021-1049
64. Roots BI: Neurofilament accumulation induced in synapses by leupeptin. Science 221:971-972, 1983
65. Nixon RA, Brown BA, Marotta CA: Limited proteolytic modification of a neurofilament protein involves a proteinase activated by endogenous levels of calcium. Brain Res 275:384-388, 1983
66. Hamakubo T, Kannagi R, Murachi T, et al: Distribution of calpains I and II in rat brain. J Neurosci 6:3103-3111, 1986
67. Nixon RA, Marotta CA: Degradation of neurofilament proteins by purified human brain cathepsin D. J Neurochem 43:507-516, 1984
68. Hanukoglu I, Fuchs E: The cDNA sequence of a human epidermal keratin: divergence of sequence but conservation of structure among intermediate filament proteins. Cell 31:243-252, 1982
69. Hanukoglu I, Fuchs E: The cDNA sequence of a type II cytoskeletal keratin reveals constant and variable structural domains among keratins. Cell 33:915-924, 1983
70. Quax W, Egberts WV, Hendricks W, et al: The structure of the vimentin gene. Cell 35:215-223, 1983
71. Lewis SA, Balcarek JM, Krek V, et al: Sequence of a cDNA clone encoding mouse glial fibrillary acidic protein: Structural conservation of intermediate filaments. Proc Natl Acad Sci USA 81:2743-2746, 1984
72. Quax W, van den Broek L, Egberts WV, et al: Characterization of the hamster desmin gene: Expression and formation of desmin filaments in nonmuscle cells after gene transfer. Cell 43:327-338, 1985
73. Steinert PM, Parry DAD, Racoosin EL, et al: The complete cDNA and deduced amino acid sequence of a Type II mouse epidermal keratin of 60,000 Da: Analysis of sequence differences between Type I and Type II keratins. Proc Natl Acad Sci USA 81:5709-5713, 1984
74. Zehner ZE, Paterson BM: Characterization of the chicken vimentin gene: single copy gene producing multiple mRNAs. Proc Natl Acad Sci USA 80:911-915, 1983
75. Geisler N, Plessmann U, Weber K: The complete amino acid sequence of the major mammalian neurofilament protein (NF-L). FEBS Letters 182:475-478, 1985
76. Lewis SA, Cowan NJ: Genetics, evolution, and expression of the 68,000-mol-wt neurofilament protein: Isolation of a cloned cDNA probe. J Cell Biol 100:843-850, 1985

77. Julien JP, Ramachandran K, Grosveld F: Cloning of a cDNA encoding the smallest neurofilament protein from the rat. Biochim Biophys Acta 825:398-404, 1985
78. Lieberburg I, Lappin I, Davies P, et al: Molecular cloning of a cDNA coding for a high molecular weight neurofilament peptide. J Cell Biol 101:17a, 1985
79. Robinson PA, Wion D, Anderton BH: Isolation of a cDNA for the rat heavy neurofilament polypeptide (NF-H). FEBS Letters 209:203-205, 1986
80. Lewis SA, Cowan NJ: Anomalous placement of introns in a member of the intermediate filament multigene family: An evolutionary conundrum. Mol Cell Biol 6:1529-1534, 1986
81. Levy E, Liem RKH, D'Eustachio P, et al: Structure and evolutionary origin of the gene encoding mouse NF-M, the middle molecular weight neurofilament protein. Eur J Biochem in press
82. Trojanowski JQ, Walkenstein N, Lee VMY: Expression of neurofilament subunits in neurons of the central and peripheral nervous system: An immunohistochemical study with monoclonal antibodies. J Neurosci 6:650-660, 1986
83. Sharp GA, Shaw G, Weber K: Immunoelectronmicroscopical localization of the three neurofilament triplet proteins along neurofilaments of cultured dorsal root ganglion neurones. Exp Cell Res 137:403-413, 1982
84. Lawson SN, Harper AA, Harper EI, et al: A monoclonal antibody against neurofilament protein specically labels a subpopulation of rat sensory neurons. J Comp Neurol 228:263-272, 1984
85. Kai-Kai MA, Anderton BH, Keen P: A quantitative analysis of the interrelationships between subpopulations of rat sensory neurons containing arginine vasopressin or oxytocin and those containing substance P, fluoride-resistant acid phosphatase or neurofilament protein. Neuroscience 18:475-486, 1986
86. Drake PF, Lasck RJ: Regional differences in the neuronal cytoskeleton. J Neurosci 4:1173-1186, 1984
87. Shaw G, Weber K: The distribution of the neurofilament triplet protein within individual neurones. Exp Cell Res 136:119-125, 1981
88. Dahl D: Immunohistochemical differences between neurofilaments in perikarya, dendrites and axons. Exp Cell Res 149:397-408, 1983
89. Bennett GS, Tapscott SJ, DiLullo, C, et al: Differential binding of antibodies against the neurofilament triplet proteins in different avian neurons. Brain Res 304:291-302, 1984
90. Lee, VMY, Carden MJ, Trojanowski JQ: Novel monoclonal antibodies provide evidence for the in situ existence of a nonphosphorylated form of the largest neurofilament subunit. J Neurosci 6:850-858, 1986
91. Tapscott SJ, Bennett GS, Toyama Y, et al: Intermediate filament proteins in the developing chick spinal cord. Dev Biol 86:40-54, 1981
92. Raju T, Bignami A, Dahl D: In vivo and in vitro differentiation of neurons and astrocytes in the rat embryo. Dev Biol 85:344-357, 1981
93. Jacobs M, Choo QL, Thomas C: Vimentin and 70K neurofilament protein co-exist in embryonic neurones from spinal ganglia. J Neurochem 38:969-977, 1982

94. Shaw G, Weber K: Differential expression of neurofilament triplet proteins in brain development. Nature 298:277-279, 1982
95. Shaw G, Weber K: The structure and development of the rat retina; an immunofluorescence microscopical study using antibodies specific for intermediate filament proteins. Eur J Cell Biol 30:219-232, 1983
96. Cochard P, Paulin D: Initial expression of neurofilaments and vimentin in the central and peripheral nervous system of the mouse embryo in vivo. J Neurosci 4:2080-2094, 1984
97. Pachter JS, Liem RKH: The differential appearance of neurofilament triplet polypeptides in the developing rat optic nerve. Dev Biol 103:200-210, 1984
98. Drager UC: Coexistence of neurofilaments and vimentin in a neurone of adult mouse retina. Nature 303:169-172, 1983
99. Shaw G, Weber K: The intermediate filament complement of the retina: A comparison between different mammalian species. Eur J Cell Biol 33:95-104, 1984
100. Bader MF, Georges E, Mushynski WE, et al: Neurofilament proteins in cultured chromaffin cells. J Neurochem 43:1180-1193, 1984
101. Granger BL, Lazarides E: Expression of the major neurofilament subunit in chicken erythrocytes. Science 221:553-556, 1983
102. Bennett GS, DiLullo C: Transient expression of a neurofilament protein by replicating neuroepithelial cells of the embryonic chick brain. Dev Biol 107:107-127, 1985
103. Osborn M, Altmannsberger M, Shaw G, et al: Various sympathetic derived human tumors differ in neurofilament expression. Virchows Arch B 40:141-156, 1982
104. Trojanowski JQ, Lee VMY, Schlaepfer WW: An immunohistochemical study of human central and peripheral nervous system tumors, using monoclonal antibodies against neurofilaments and glial filaments. Human Pathol 15:248-257, 1984
105. Dahl D, Bignami A: Neurofilament protein in clonal lines of mouse neuroblastoma. Dev Neurosci 1:142-152, 1978
106. Dahl D, Rustigian R, Bignami A: Expression of the 70 Kdalton neurofilament protein in clonal lines of mouse neuroblastoma. Brain Res 299:354-357, 1984
107. Lee V, Trojanowski JQ, Schlaepfer WW: Induction of neurofilament triplet proteins in PC12 cells by nerve growth factor. Brain Res 238:169-180, 1982
108. Lee VMY, Page C: The dynamics of nerve growth factor-induced neurofilaments and vimentin filament expression and organization in PC12 cells. J Neurosci 4:1705-1714, 1984
109. Lee VMY, Andrews PW: Differentiation of NTERA-2 clonal human embryonal carcinoma cells into neurons involves the induction of all three neurofilament proteins. J Neurosci 6:514-521, 1986
110. Broers JLV, Carney DN, Rot MK, et al: Intermediate filament proteins in classic and variant types of small cell lung carcinoma cell lines: A biochemical and immunochemical analysis using a panel of monoclonal and polyclonal antibodies. J Cell Sci 83:37-60, 1986

111. Eagles PAM, Gilbert S, Maggs A: The polypeptide composition of axoplasm and of neurofilaments from the marine worm Myxicola infundibulum. Biochem J 199:89-100, 1981
112. Shaw G, Debus E, Weber, K: The immunological relatedness of neurofilament proteins of higher vertebrates. Eur J Cell Biol 34:130-138, 1984
113. Lasek RJ, Phillips L, Katz MJ, et al: Function and evolution of neurofilament proteins. Ann NY Acad Sci 455:462-478, 1985
114. Gambetti P, Autilio-Gambetti L, Papasozomenos SC: Bodian's silver method stains neurofilament polypeptides. Science 213:1521-1522, 1981
115. Phillips LL, Autilio-Gambetti L, Lasek RJ: Bodian's silver method reveals molecular variation in the evolution of neurofilament proteins. Brain Res 278:219-223, 1983
116. Grafstein B, Forman DS: Intracellular transport in neurons. Physiol Reviews 60:1167-1283, 1980
117. Lasek RJ, Garner JA, Brady ST: Axonal transport of the cytoplasmic matrix. J Cell Biol 99:212s-221s, 1984
118. Black MM, Lasek RJ: Slow components of axonal transport: Two cytoskeletal networks. J Cell Biol 86:616-623, 1980
119. Hoffman PN, Lasek RJ: The slow component of axonal transport: Identification of major structural polypeptides of the axon and their generality among mammalian neurons. J Cell Biol 66:351-366, 1975
120. McQuarrie IG, Brady ST, Lasek RJ: Diversity in the axonal transport of structural proteins: Major differences between optic and spinal axons in the rat. J Neurosci 6:1593-1605, 1986
121. Oblinger MM, Brady ST, McQuarrie IG, et al: Cytotypic differences in the protein composition of the axonally transported cytoskeleton in mammalian neurons. J Neurosci 7:453-462, 1987
122. Allen RD, Weiss DG, Hayden JH, et al: Gliding movement of and bidirectional transport along single native microtubules from squid axoplasm: Evidence for an active role of microtubules in cytoplasmic transport. J Cell Biol 100:1736-1752, 1985
123. Vale RD, Reese TS, Sheetz MP: Identification of a novel force-generating protein, kinesin, involved in microtubule-based motility. Cell 42:39-50, 1985
124. Papasozomenos SC, Yoon M, Crane R, et al: Redistribution of protein of fast axonal transport following adminstration of β,β'-iminodipropionitrile: A quantitative autoradiographic study. J Cell Biol 95:672-675, 1982
125. Griffin JW, Fahnestock KE, Price, et al: Microtubule-neurofilamnet segregation produced by β,β'-iminodipropionitrile: Evidence for the association of fast axon transport with microtubules. J Neurosci 3:557-566, 1983
126. Hirokawa N, Bloom GS, Vallee RB: Cytoskeletal architecture and immunocytochemical localization of microtubule-associated protein in regions of axons associated with rapid axonal transport: The β,β'-iminodipropionitrile-intoxicated axon as a model system. J Cell Biol 101:227-239, 1985

127. Hendrickson AE, Cowan WM: Changes in the rate of axoplasmic transport during postnatal development of the rabbit's optic nerve and tract. Exp Neurol 30:403-422, 1971
128. Willard M, Simon C: Modulations of neurofilament axonal transport during the development of rabbit retinal ganglion cells. Cell 35:551-559, 1983
129. Nixon RA, Logvinenko KB: Multiple fates of newly synthesized neurofilament proteins: Evidence for a stationary neurofilament network distributed nonuniformly along axons of retinal ganglion cell neurons. J Cell Biol 102:647-659, 1986
130. Friede RL, Samorajski T: Axon caliber related to neurofilaments and microtubules in sciatic nerve fibers of rats and mice. Anat Rec 167:379-388, 1970
131. Friede RL: Changes in microtubules and neurofilaments in constricted, hypoplastic nerve fibers. Acta Neuropath (Berl) Suppl 5:216-225, 1971
132. Hoffman PN, Griffin JW, Price DL: Contol of axonal caliber by neurofilament transport. J Cell Biol 99:705-714, 1984
133. Hoffman PN, Thompson GW, Griffin JW, et al: Changes in neurofilament transport coincide temporally with alterations in the caliber of axons in regenerating motor fibers. J Cell Biol 101:1332-1340, 1985
134. Nukada H, Dyck PJ, Low PA, et al: Axonal caliber and neurofilaments are proportionately decreased in galactose neuropathy. J Neuropath Exp Neurol 45:140-150, 1986
135. Nukada H, Dyck PJ: Decreased axon caliber and neurofilaments in hereditary motor and sensory neuropathy, type I. Ann Neurol 16:238-241, 1984
136. Price DL: New perspectives on Alzheimer's disease. Annu Rev Neurosci 9:489-512, 1986
137. Goldman JE, Yen SH, Chiu FC, et al: Lewy bodies of Parkinson's disease contain neurofilament antigens. Science 221:1082-1084, 1983
138. Hirano A, Donnenfeld H, Sasaki S et al: Fine structural observations of neurofilamentous changes in amyotrophic lateral sclerosis. J Neuropathol Exp Neurol 43:461-470, 1984
139. Hirano A, Nakano I, Kurland LT, et al: Fine structural study of neurofibrillary changes in a family with amyotrophic lateral sclerosis. J Neuropathol Exp Neurol 43:471-480, 1984
140. Perl DP, Gajdusek DC, Garruto RM, et al: Intraneuronal aluminum accumulation in amyotrophic lateral sclerosis and parkinsonism-dementia of Guam. Science 217:1053-1055, 1982
141. Anderton BH, Breinburg D, Downes MJ, et al: Monoclonal antibodies show that neurofibrillary tangles and neurofilaments share antigenic determinants. Nature 298:84-86, 1982
142. Rasool CG, Selkoe DJ: Sharing of specific antigens by degenerating neurons in Pick's disease. New Engl J Med 312:700- 705, 1985
143. Peiffer J, Scholte W, Bishoff A, et al: Generalized giant axonal neuropathy - a filament-forming disease of neuronal, endothelial, glial and Schwaan cells in a patient without kinky hair. Acta Neuropathol 40:213-218, 1977

144. Yen SHC, Gaskin F, Terry RD: Immunocytochemical studies of neurofibrillary tangles. Am J Pathol 104:77-89, 1981
145. Dahl D, Selkoe DJ, Pero RT, et al: Immunostaining of neurofibrillary tangles in Alzheimer's senile dementia with a neurofilament antiserum. J Neurosci 2:113-119, 1982
146. Gambetti P, Shecket G, Ghetti B, et al: Neurofibrillary changes in human brain: An immunocytochemical study with a neurofilament antiserum. J Neuropathol Exp Neurol 42:69-79, 1983
147. Cork LC, Sternberger NH, Sternberger LA, et al: Phosphorylated neurofilament antigens in neurofibrillary tangles in Alzheimer's disease. J Neuropathol Exp Neurol 45:56-64, 1986
148. Wischik CM, Crowther RA, Stewart M, et al: Subunit structure of paired helical filaments in Alzheimer's disease. J Cell Biol 100:1905-1912, 1985
149. Selkoe DJ, Ihara Y, Salazar FJ: Alzheimer's disease: Insolubility of partially purified paired helical filaments in sodium dodecyl sulfate and urea. Science 215:1243-1245, 1982
150. Perry G, Rizzuto N, Autilio-Gambetti L, et al: Paired helical filaments from Alzheimer disease patients contain cytoskeletal components. Proc Natl Acad Sci USA 82:3916-3920, 1985
151. Kosik KS, Duffy LK, Dowling MM, et al: Microtubule-associated protein 2: Monoclonal antibodies demonstrate the selective incorporation of certain epitopes into Alzheimer neurofibrillary tangles. Proc Natl Acad Sci USA 81:7941-7945, 1984
152. Grundke-Iqbal I, Iqbal K, Tung YC, et al: Abnormal phosphorylation of the microtubule-associated protein τ (tau) in Alzheimer cytoskeletal pathology. Proc Natl Acad Sci USA 83:4913-4917, 1986
153. Ihara Y, Nukina N, Miura R, et al: Phosphorylated tau protein is integrated into paired helical filaments in Alzheimer's disease. J Biochem 99:1807-1810, 1986
154. Griffin JW, Price DL, Hoffman PN: Neurotoxic probes of the axonal cytoskeleton. Trends in Neurosci December 1983, pp 490-495
155. Bisby MA, Redshaw JD: Acrylamide neuropathy: Changes in the composition of proteins of fast axonal transport resemble those observed in regenerating axons. J Neurochem 48:924-928, 1987
156. Monaco S, Autilio-Gambetti L, Zabel D, et al: Giant axonal neuropathy: Acceleration of neurofilament transport in optic axons. Proc Natl Acad Sci USA 82:920-924, 1985
157. Troncoso JC, Hoffman PN, Griffin JW, et al: Aluminum intoxication: A disorder of neurofilament transport in motor neurons. Brain Res 342:172-175, 1985
158. Hoffman PN, Clark AW, Carroll PT, et al: Slow axonal transport of neurofilament proteins: Impairment by , β,β'-iminodipropionitrile administration. Science 202:633-635, 1978
159. Gajdusek DC: Hypothesis: Interference with axonal transport of neurofilaments as a common pathogenetic mechanism in certain diseases of the central nervous system. New Engl J Med 312:714- 719, 1985

160. Graham DG, Szakal-Quin G, Priest JW, et al: In vitro evidence that covalent crosslinking of neurofilaments occurs in γ-diketone neuropathy. Proc Natl Acad Sci USA 81:4979-4982, 1984
161. Lapadula DM, Irwin RD, Suwita E, et al: Cross-linking of neurofilament proteins of rat spinal cord in vivo after administration of 2,5-hexandione. J Neurochem 46:1843-1850, 1986
162. Sotelo J, Gibbs CJ, Gajdusek DL: Autoantibodies against axonal neurofilaments in patients with kuru and Creutzfeldt-Jakob disease. Science 210:190-193, 1980
163. Bahmanyar S, Liem RKH, Griffin JW, et al: Characterization of antineurofilament autoantibodies in Creutfeldt-Jakob disease. J Neuropathol Exp Neurol 43:369-375, 1984
164. Toh BH, Gibbs CJ, Gajdusek DC, et al: The 200- and 150-kDa neurofilament protein react with IgG autoantibodies from chimpanzees with kuru or Creutzfeldt-Jakob disease; a 62-kDa neurofilament-associated protein reacts with sera from sheep with natural scrapie. Proc Natl Acad Sci USA 82:3894-3896, 1985

THE PHOSPHORYLATION OF THE MICROTUBULE-ASSOCIATED TAU PROTEINS

Jacques Baudier and R. David Cole

Department of Biochemistry
University of California
Berkeley, CA 94720
(415) 642-7178

INTRODUCTION

Extracellular signals, including a variety of neurotransmitters and hormones produce diverse physiological responses in neurons but until recently, little was known about the molecular mechanims that underlie these responses. There is now considerable evidence indicating that protein phosphorylation plays an important role in neuronal responses to extracellular messengers (for review see [1]). Protein phosphorylation is involved in a shuttle: i) the substrate protein is phosphorylated when a protein kinase transfers the terminal phosphate from ATP to the substrate protein; the addition of the charged phosphate group to the protein modulates its function, presumably by changing its conformation; ii) the phosphate group can be removed from the protein by a phosphatase restoring the protein to its original functional state. Such a shuttle is now evident in microtubule proteins. Microtubules are one of the major constituents of neuronal cytoskeletons and play important roles in regulating cell morphology, intracellular transport processes and secretions. The microtubules are polymers of tubulin and they also contain accessory proteins, notably MAP1, MAP2, and the tau proteins.[2]

The tau proteins have been purified from brain of several mammalian species. In bovine brain , they are a family of four closely related polypeptides, as indicated by their similar peptide maps and amino acid compositions, but they differ in their apparent molecular weights (Mr = 55,000-65,000) as determined by SDS-polyacrylamide gel electrophoresis (3). Recently, it was reported that phosphorylated tau proteins are the major antigenic component of the paired helical filaments that characterize degenerative human neurons in Alzheimer's disease.[4,5,6] The enrichment in phosphorylated tau proteins seemed to be accompanied by an impoverishment in microtubules. It is significant, then, that Lindwall and Cole[7] found that phosphorylation of tau reduced its ability to promote microtubule assembly. Moreover, at high ratios of phosphorylated tau to tubulin, anomalies were reported in the turbidity assay for microtubule assembly, indicating aberrations in the structure of the tubulin polymers.[7]

The mechanisms that dictate microtubule assembly, the interactions of microtubules with each other, or with membranes and their interactions with other cytoplasmic structures or proteins, are not completely understood. However, on the basis of recent available data, it became evident that phosphorylation processes of tubulin[8] and the microtubule-associated proteins, MAP-2 and tau are major molecular events that might regulate microtubule properties and functions.[7,9,10] Such an assumption is supported by the recent finding that tau proteins are phosphoproteins.[11-13] Apparently, the original isolation of tau proteins[2] produced tau in a particular state of phosphorylation because it allowed phosphorylation to occur during the isolation process.[11] However, a quick, direct purification technique showed that tau existed in multiple phosphorylation states *in vivo*.[11] Tau proteins associated with paired helical filaments in Alzheimer brain neurons have been found to be highly phosphorylated.[12,13] There is therefore a considerable interest in understanding the factors that might regulate tau phosphorylation *in vivo* and thus probably their functions.

Two different modes of phosphorylation of tau protein have been characterized. Mode I phosphorylation induces a conformational change in tau protein structures that is responsible for a decrease in mobility on SDS-polyacrylamide gel electrophoresis[11] and makes a highly conserved tau epitope inaccessible to antibody to tau.[12,13] This phosphorylation was found to inhibit tau in its promotion of microtubule formation.[7] Mode II phosphorylation causes little change in the mobility of tau on SDS gels.[3,11] The primary focus of this review will be on the protein kinases that have been recently found to catalyze mode I and mode II phosphorylation of tau proteins *in vitro*.

Mode I phosphorylation of tau proteins by a Ca2+/calmodulin-dependent protein kinase

Activation of cell surface receptors by a variety of stimuli leads to an increase in intracellular Ca^{2+} concentration. Ca^{2+} acting as a "second messenger," activates several protein kinases that amplify the initial signal and may mediate many of the effects of Ca^{2+} on neuronal function. The discovery of the ubiquitous Ca^{2+}-binding protein, calmodulin (CaM), facilitated the elucidation of many effects of Ca^{2+} by finding that many enzymes are regulated directly by the Ca^{2+}-CaM complex (14). Schulman and Greengard (15) first reported the presence of a membrane-bound Ca^{2+}-calmodulin-dependent protein kinase in brain and Fukunaga et al.[16] subsequently purified a cytosolic form of the enzyme. To date, several laboratories have purified Ca^{2+}-calmodulin-dependent protein kinases. The enzymes have been purified using myosin light chain[16], tryptophan hydroxylase[17], synapsin[18], tubulin [8,19], microtubule associated MAP-2 protein [19,20], and casein[21] as exogenous substrate. All preparations, except the enzyme purified as a myosin light-chain kinase showed a similar subunit composition with a molecular weight of the holoenzymes varying from 350,000 to 650,000. Although major differences in substrate specificity have been reported it is not yet possible to determine whether these represent the same enzyme that has been prepared and assayed under different conditions or whether there is a family of related enzymes. Immunohistochemical studies on the location of the enzyme showed that it is distributed throughout the brain and that it is mainly concentrated in neurons.[22] In 1984, Schulman[20] first reported that the Ca^{2+}-CaM-dependent MAP-2 kinase purified from rat brain may also use tau protein as a substrate. Preliminary experiments in our laboratory using the kinase kindly given to us by Schulman showed that it did indeed phosphorylate purified tau protein. A similar enzyme was prepared from bovine forebrain according to the method of McGuinness _et al._[23] with minor modifications. The final step consisted of a calmodulin-Sepharose affinity chromatography.

Phosphorylation of tau protein by the enzyme we prepared proved to induce shifts in the mobilities of the phosphorylated tau protein that characterize mode I phosphorylation (Fig. 1, upper panel). Furthermore, phosphorylation of tau by this kinase was almost totally dependent on the presence of calcium and calmodulin. There was no apparent preference of the kinase for one or another tau protein species, since they all seemed to be phosphorylated by the kinase to the same extent (Fig. 1, lower panel, lane 5). Further studies on individually separated tau species has allowed the demonstration that each tau protein species incorporated one phosphate group per mole of tau and was shifted to a single slower migrating electrophoretic band as a result of phosphorylation (Fig. 1).

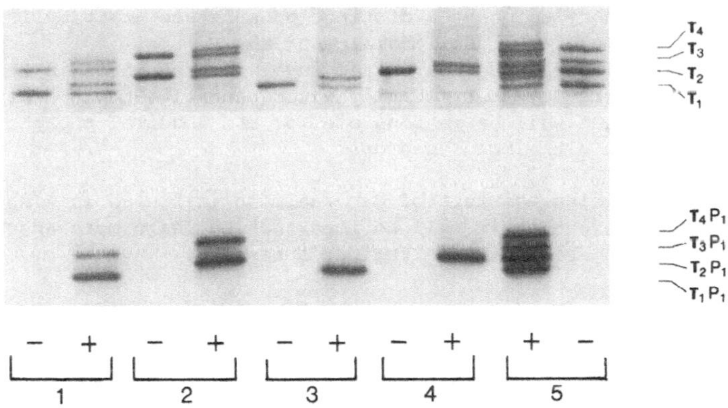

Figure 1. Coomassie blue staining (upper panel) and autoradiogram (lower panel) showing the different peptide and phosphopeptide patterns observed after treatment of various dephosphorylated bovine brain tau protein combinations with the bovine Ca^{2+}/calmodulin-dependent protein kinase. Lane 1: mixture of τ_1 and τ_3; Lane 2: mixture of τ_2 and τ_4; Lane 3: τ_1; Lane 4: τ_2; Lane 5: crude τ proteins. The samples containing the Ca^{2+}/CaM kinase (+) were incubated for 20 min at 30°C with 20 mM Tris-HCl, pH 7.5, 36 μM [γ-^{32}P]ATP, 18 mM $MgCl_2$ and 30 μl (about 40 μg/ml) of the purified kinase in a total volume of 55 μl in the presence of 1.5 μM calmodulin and 45 μM calcium. The reference samples (-) were incubated in Tris-HCl buffer. The reaction was stopped by adding 15 μl of SDS buffer and boiled for 5 min. Samples were then electrophoresed on 10% SDS-polyacrylamide gels. The gel was stained with Coomassie blue and dried before autoradiography. The upper and lower panels represent,

respectively, the Coomassie blue staining and autoradiography of the same gel. The different tau protein species are numbered T_1 to T_4 and P_1 refers to mode I phosphorylated tau protein. The method for separation of the different tau protein species will be described elsewhere.

For a number of reasons, we believe that there is physiological significance in the phosphorylation of tau by the Ca^{2+}/calmodulin-dependent kinase. First, the mode of phosphorylation produced by this kinase is the one most clearly shown to exist *in vivo*.[11-13] Second, the phosphorylation by the Ca^{2+}/calmodulin-dependent kinase appears to produce a conformational change in tau and it seems likely that such a change would modulate the functional properties of tau.[7] Finally, the possibility of physiological significance in tau phosphorylation by the Ca^{2+}/calmodulin kinase is enhanced by the recent reports that tau proteins in Alzheimer neurons are predominantly in the form produced by mode I phosphorylation.[12,13] Concerning this last point, a major challenge in understanding the biochemical alteration in Alzheimer neurons seems now to be elucidation of the molecular events that underlie such longlasting changes in the phosphorylation of tau proteins. The Ca^{2+}/calmodulin-dependent kinase is one enzyme that could mediate such changes because it appears to act as a molecular switch that remains active long after an initial triggering event.[24,25] Ca^{2+} and calmodulin stimulate autophosphorylation of the enzyme and incorporation of 3-12 phosphate groups per holoenzyme.[24] The phosphorylated form which results from autocatalysis becomes autonomous, being completely independent of Ca^{2+} and/or calmodulin. Such properties allow the enzyme to remain active long after the decay of an initial Ca^{2+} signal. Dephosphorylation by a distinct phosphatase activity is required to turn the enzyme off.[24] An imbalance of the autophosphorylation-phosphatase system involving reduction in the level of phosphatase activity that normally would control autophosphorylation of the enzyme might well be responsible for the abnormal phosphorylation of tau protein in Alzheimer's neurons.

To better define the role of Ca^{2+}/calmodulin kinase in Alzheimers disease, we believe that it will be important to learn more about the regulation of its autonomy in Alzheimer's neurons.

Mode II phosphorylation of tau proteins

It has been shown that purified tau proteins can exist in a second mode of phosphorylation, mode II, which hardly changes the mobility of tau on SDS-polyacrylamide gel electrophoresis.[3,11] *In vitro* studies of tau phosphorylation by the Ca^{2+}/phospholipid-dependent protein kinase C demonstrated that tau that had been dephosphorylated with alklaline phosphatase could be rephosphorylated by this kinase (Figure 2) without an appreciable shift in mobility.[26] However, a sub-stoichiometric incorporation of radioactive phosphate group per mol of tau protein was consistently observed (0.3 - 0.6 mole $PO_4^{=}$ / mole tau protein), independently of the kinase C concentration used (unpublished). We have since learned that the phosphoryl group introduced into tau by kinase C is unusually resistant to removal by alkaline phosphatase. Thus, the low stoichiometry measured by radioactive phosphate incorporation was due to the fact that the substrate tau was already partially phosphorylated with non-radioactive isotope.

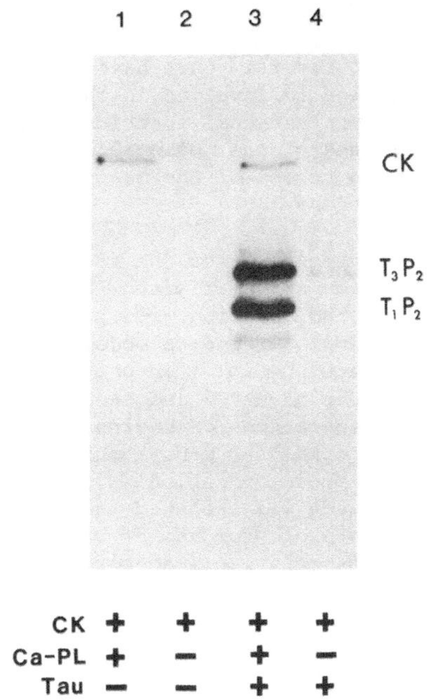

Figure 2. Calcium/phospholipid-dependent phosphorylation of bovine brain tau protein by protein kinase C. A purified tau protein preparation that contained mainly two tau protein species, T_2 and τ_3 was used. Each lane of the autoradiograms represents a reaction mixture of protein kinase C in 20 mM Tris, pH 7.4, with the additions indicated below the autoradiograms. The additions were in the following final concentrations: total Ca^{2+}, 1.2 mM; phospholipids, 246 μM phosphatidyl serine, 7 μM diolein; tau, 2 μM. Protein kinase C was purified from rat brain (a generous gift of Drs. Mochly-Rosen and D.E. Koshland, Jr.). The phosphorylated polypeptide in lanes 1 and 3 represents autophosphorylated forms of protein kinase C (CK). The different phosphorylated tau protein species are numbered T_1 and T_3, and P_2 refers to mode II phosphorylated tau protein.

Like the Ca^{2+}/calmodulin-dependent kinase, protein kinase C has a relatively broad substrate specificity *in vitro*, but there is less evidence than in the case of Ca^{2+}/calmodulin-dependent kinase that the observed phosphorylation of tau by kinase C might correspond to mode II phosphorylation that occurs *in vivo*. Further studies on the phosphorylation of tau protein by other protein kinases and comparison of peptide maps of the *in vivo* and *in vitro* phosphorylated tau proteins might help to identify the kinases that catalyze mode II phosphorylation of tau *in vivo*.

Conclusion

Studies over the past several years have provided direct evidence that protein phosphorylation is involved in the regulation of neuronal function. A major factor in neuronal function could be microtubules which are major constituents of the neuronal cytoskeleton, and phosphorylation appears to be one of the major molecular mechanisms that regulate microtubule functions.

Only a beginning has been made on the identification of the kinases involved in microtubule regulation *in vivo*. Three distinct classes of protein kinase are under consideration, i.e., cyclic AMP-dependent protein kinase, protein kinase C, and calmodulin dependent kinase II. All of these kinases mediate the *in vitro* phosphorylation of tubulin and microtubule associated proteins MAP-2 and tau. Lindwall and Cole[11] reported that in crude preparations of bovine brain microtubule proteins, a kinase was present that catalysed mode I phosphorylation and another that accomplished mode II phosphorylation. Since the Ca^{2+}-calmodulin dependent kinase in our study was isolated from bovine brain, it seems likely to be the one responsible for the *in vivo* phosphorylation of tau, such as that observed in quickly isolated normal tau preparations, and in the paired helical filaments of Alzheimer's neurons. The protein kinase C that was used in our work was isolated from rat brain and it is therefore reasonable to suppose that tau is exposed to this enzyme *in vivo*. However, it must be kept in mind that because electrophoretic mobility does not signal mode II phosphorylation, as it does mode I, it has not yet been proven directly that tau *in vivo*, or even in freshly isolated preparations is partially or completely phosphorylated in mode II. Despite this caveat, it seems likely to us that mode II phosphorylation, as well as mode I, occurs *in vivo*. It will be interesting to learn whether or not there are any interdependencies between the two different modes of phosphorylation; that is, does mode I phosphorylation regulate mode II or vice versa? It also remains to elucidate how these two different modes of phosphorylation can affect tau protein structure and function. A number of functional consequences of tau phosphorylation could be postulated. It is already known[7] that the ability of tau to promote microtubule assembly is reduced by mode I phosphorylation, but not eliminated. It might be asked, then, whether microtubules assembled with phosphorylated tau proteins are more or less stable toward, for example, calcium, than are microtubules assembled with non-phosphorylated tau. Further, the phosphorylation of tau may affect interactions, perhaps even when it is incorporated into microtubules. For example, it was reported[26] that phosphorylation (unidentified mode) decreased the effect of tau on actin gelation. Whatever particular effects are ultimately found to be relevant, the occurrence of two modes of phosphorylation along with the existence of substantial amounts of nonphosphorylated tau *in vivo*, point toward a dynamic regulatory system with diverse effects.

References

1. M.D. Browning, R. Huganir, and P. Greengard, Protein phosphorylation and neuronal function. J. Neurochem. 45: 11-23 (1985).
2. M.D.Weingarten, A protein factor essential for microtubule assembly. Proc. Natl. Acad. Sci. USA 72: 1858-1862 (1975).
3. D.W. Cleveland, S.Y. Hwo, and M.W Kirschner, Physical and chemical properties of purified tau factor and the role of tau in microtubule assembly. J. Mol. Biol. 116: 227-247 (1977).

4. N. Nukina and Y. Ihara, One of the antigenic determinants of paired helical filaments is related to tau protein. J. Biochem. 99: 1541-1544 (1986).
5. I. Grundke-Iqbal, et al., Microtubule-associated protein tau. A component of Alzheimer paired helical filaments. J. Biol. Chem. 261: 6084-6089 (1986).
6. K.S. Kosik, C.L. Joachim, and D.J. Selkoe, Microtubule-associated protein tau (τ) is a major antigenic component of paired helical filaments in Alzheimers disease. Proc. Natl. Acad. Sci. USA 83: 4044-4048 (1986).
7. G.Lindwall and R.D. Cole, Phosphorylation affects the ability of tau protein to promote microtubule assembly. J. Biol. Chem. 259: 5301-5305 (1984).
8. F. Wandosell, et al., Phosphorylation of tubulin by a calmodulin-dependent protein kinase. J. Biol. Chem. 261: 10332-10339 (1986).
9. A.S.N. Murthy, G.T. Bramblett, and M. Flavin, The sites at which brain microtubule-associated protein 2 is phosphorylated in vivo differ from those accessible to cAMP-dependent kinase in vitro. J. Biol. Chem. 260: 4363-4370 (1985).
10. S.C. Selden, and T.D. Pollard, Phosphorylation of microtubule-associated proteins regulates their interaction with actin filaments. J. Biol. Chem. 258: 7064-7071 (1983).
11. G. Lindwall, and R.D. Cole, The purification of tau protein and the occurrence of two phosphorylation states of tau in brain. J. Biol. Chem. 259: 12241-12245 (1984).
12. I. Grundke-Iqbal et al., Abnormal phosphorylation of the microtuble-associated protein τ (tau) in Alzheimer cytoskeletal pathology. Proc. Natl. Acad. Sci. USA 83: 4913-4917 (1986).
13. Y. Ihara, et al., Phosphorylated tau protein is integrated into paired helical filament in Alzheimers disease. J. Biochem. 99: 1807-1810 (1986).
14. H. Schulman, Calcium-dependent protein kinases and neuronal function. TIBS, May 1984, pp. 188-192 (1984).
15. H. Schulman, and P. Greengard, Stimulation of brain membrane protein phosphorylation by calcium and an endogenous heat-stable protein. Nature 271: 478-479 (1978).
16. K. Fukunaga, et al., Purification and characterization of a Ca^{2+}- and calmodulin-dependent protein kinase from rat brain. J. Neurochem. 39: 1607-1617 (1982).
17. T. Yamauchi and H. Fujisawa, Purification and characterization of the brain calmodulin-dependent protein kinase (kinase II) which is involved in the activation of tryptophan 5-monooxygenase. Eur. J. Biochem. 132: 15-21 (1983).
18. M.K. Bennett, N.E. Erondu, and M.B. Kennedy, Purification and characterization of a calmodulin-dependent protein kinase that is highly concentrated in brain. J. Biol. Chem. 258: 12735-12744 (1983).
19. J.R. Goldenring, et al., Purification and characterization of a calmodulin-dependent kinase from rat brain cytosol is able to phosphorylate tubulin and microtubule-associated protein. J. Biol. Chem. 258: 12632-12640 (1983).
20. H. Schulman, Phosphorylation of microtubule-associated proteins by a Ca^{2+}/calmodulin-dependent protein kinase. J. Cell Biol. 99: 11-19 (1984).
21. J. Kuret and H. Schulman, Purification and characterization of a Ca^{2+}/calmodulin-dependent protein kinase from rat brain. Biochemistry 23: 5495-5504 (1984).

22. C.C. Ouimet, T.L. McGuinness, and P. Greengard, Immunocytochemical localization of calcium/calmodulin-dependent protein kinase II in rat brain. Proc. Natl. Acad. Sci. USA 81: 5604-5608 (1984).
23. T.L. McGuinness, Y. Lai, and P. Greengard, Ca^{2+}/calmodulin-dependent protein kinase II. Isozymic forms from rat forebrain and cerebellum. J. Biol. Chem. 260: 1697-1704 (1985).
24. S.G. Miller, and M.B. Kennedy, Regulation of brain type II Ca^{2+}/calmodulin-dependent protein kinase by autophosphorylation: A Ca^{2+}-triggered molecular switch. Cell 44: 861-870 (1986).
25. L.L. Lou, S.J. Lloyd, and H. Schulman, Activation of the multifunctional Ca^{2+}/calmodulin-dependent protein kinase by autophosphorylation: ATP modulates production of an autonomous enzyme. Proc. Natl. Acad. Sci. USA 83: 9497-9501 (1986).
26. J. Baudier, et al, A comparison of S100b protein with calmodulin: Interactions with melittin and microtubule associated tau proteins: Inhibition of the phosphorylation of tau protein by protein kinase C. Biochemistry, in press (1987).

HUMAN AMYLOIDOSIS AND IN VITRO FORMATION OF ALZHEIMER AMYLOID FIBRILS

Eduardo M. Castaño and Blas Frangione*

Department of Pathology and *Kaplan Cancer Center
New York University Medical Center, 550 First Avenue
New York, N.Y. 10016

Amyloidosis is a generic term that describes the chronic deposition in the extracellular and intracellular spaces of abnormal fibrillar proteins. These proteins share the properties of a β-pleated sheet secondary structure, insolubility, green birefringence after Congo red staining and the formation of fibrils with a typical electron microscopic appearance.[1,2] The deposition of amyloid fibrils is the resultant of a diversity of pathological conditions, and the localization of these deposits is probably determined by tissue-related factors as well as by the type of protein involved (Table 1). Primary amyloidosis presents in 90% of the cases an M component that can be detected in serum, urine or cerebrospinal fluid.[3,4] This observation and the presence of multiple myeloma in 20% of the cases, suggest that a plasma cell dysfunction plays an important role in its pathogenesis.[5] The amyloid protein in primary amyloidosis has been shown to be the amino terminal fragment of immunoglobulin light chains with a variable molecular weight, ranging from 5-23 kDa.[6,7] Some subgroups of light chains appear to have a highly amyloidogenic primary structure, as shown by preferential association of λVI with primary amyloidosis and the in vitro formation of amyloid fibrils from some, but not all, of the Bence Jones proteins studied.[8-10] Secondary amyloidosis is related to acquired chronic inflammatory conditions such as rheumatoid arthritis, tuberculosis, leprosy, bronchiectasia, and osteomyelitis. Several malignancies including Hodgkin's Disease and renal-cell carcinoma[11,12] have been related to this complication as well as the autosomal-recessive Familial Mediterranean Fever.[13] The abnormal fibrils in these cases are made up of a protein called AA. Protein AA has been shown to be the amino terminal fragment, usually with a length of 76 residues, of a serum precursor SAA.[14-16] SAA is an α2 globulin with a molecular weight of 12 kDa, part of the HDL3 complex.[17,18] It behaves as an acute phase reactant, but its physiological role remains unknown.[19] Two major isotypes of SAA have been described in humans[20] but it has not been shown thus far that any of these isotypes is preferentially associated with amyloid formation. In these two groups of systemic amyloidosis (AA and AL), proteolytic cleavage of the serum precursors by the reticuloendothelial system has been suggested as a common mechanism of amyloidogenesis.[1] Moreover, incomplete breakdown of SAA by monocytes of patients with secondary amyloidosis, releasing fragments with the electrophoretic mobility of protein AA supports this concept.[21]

The types of amyloidosis that have been related to prealbumin or transthyretin as the serum precursor are: 1) familial amyloidotic

Table 1. CLINICAL AND BIOCHEMICAL CLASSIFICATION OF HUMAN AMYLOIDOSIS

	CLINICAL TYPE	FAMILIAL	AMYLOID PROTEIN
SYSTEMIC	Primary		Immunoglobulin Light Chains (6-10)
	Myeloma-Associated		Immunoglobulin Light Chains (3,4)
	Reactive or Secondary		Protein AA (11-18)
	Familial Mediterranean Fever	Autosomal-Recessive	Protein AA (13)
	Hemodialysis-Related Amyloidosis		β_2-microglobulin (31,32)
	Familial Amyloidotic Polyneuropathy (I-II)	Autosomal-Dominant	Prealbumin Variants (24-28)
	Senile Cardiac Amyloidosis		Prealbumin (22)
LOCALIZED	Hereditary Cerebral Hemorrhage With Amyloidosis (Icelandic Type)	Autosomal-Dominant	Cystatin C Variant (35,36)
	Hereditary Cerebral Hemorrhage With Amyloidosis (Dutch Type)	Autosomal-Dominant	? (37,38)
	Alzheimer's Disease & Down's Syndrome	*	β-protein or A_4 (59-62)
	Transmissible Spongiform Encephalopathies**		Protease Resistant Protein 27-30 kDa (83)
	Medullar Carcinoma of the Thyroid		Procalcitonin (33)
	Others (Cutaneous, Pancreatic, etc.)		?

*Familial cases of Alzheimer's Disease have been described with an autosomal-dominant mode of transmission.

**Creukzfeldt-Jakob Disease, Kuru, Gerstmann-Straussler Syndrome, and Scrapie in animals.

polyneuropathies (FAP); 2) senile cardiac amyloidosis[22]; and 3) familial cardiomyopathy of Danish origin.[23] In FAP of different kindreds, several variant prealbumin genes have been found. FAP type I, or lower limb polyneuropathy, described in families of Portuguese, Japanese, and Swedish origin, presents a substitution of methionine for valine at position 30.[24-26] In the Jewish kindred (patient SKO), a prealbumin variant with substitutions of glycine for threonine in position 49 and isoleucine for phenylalanine in position 33 have been observed.[27] In FAP type II, or upper limb polyneuropathy, described in kindreds of Swiss-German origin, a serine for isoleucine in position 84 has been found.[28] It has been suggested that these mutations can make the transthyretin monomers prone to polymerization and/or specific proteolytic cleavage leading to amyloid deposition in peripheral nerves and vessel walls of several organs.[27] β_2-microglobulin is an 11.8 kDa protein associated with class 1 histocompatibility antigens in cell membranes that is normally found in the circulation.[29] High serum levels of this protein are seen in renal diseases, malignancies, and chronic inflammation.[30] Recently, it has been shown that intact normal β_2-microglobulin is the main component of the hemodialysis related amyloidosis (HRA).[31,32] This finding together with the striking predilection of HRA for bone and synovium, point at the importance of tissue-related factors involved in the deposition of β_2-microglobulin as amyloid fibrils. In addition, HRA represents an example of amyloidogenesis without proteolysis of the serum precursor as an essential mechanism.

In the group of localized amyloidosis, some of the amyloid-related proteins that have been characterized include: procalcitonin in the amyloid of the medullar carcinoma of the thyroid[33] and Cystatin C or γ-trace in patients with hereditary cerebral hemorrhage with amyloidosis (HCHWA, Icelandic type), a form of amyloid deposition restricted to leptomeningeal and cortical small arteries.[34] In this case, a substitution of glutamine for leucine in position 68 of the normal molecule has been found and is yet unclear whether this represents a normal polymorphism or an inherited point mutation with pathogenic importance.[35,36] Another type of autosomal-dominant HCHWA has been described in kindreds of Dutch origin[37,38] and the amyloid protein in these patients appears not to be immunologically related to Cystatin C (unpublished observations).

Regardless of the biochemical diversity of amyloid proteins, there are several common factors that appear to interact in the pathogenesis of most of the types of amyloidosis considered.

1) The presence of a circulating soluble precursor protein that includes an amyloidogenic primary structure. In many cases, a β-pleated proteolytic fragment, resistant to further digestion, is released from the precursor as the amyloid subunit. However, in some cases the intact molecule can be the main amyloid component as well, i.e., β_2-microglobulin in HRA and prealbumin monomers in familial polyneuropathies.

2) Elevated serum levels of the precursor, reflecting overproduction, impaired clearance or a combination of the two.

3) Abnormal processing of the soluble precursor. In some instances this can be determined by the primary structure of the precursor itself as it has been proposed for prealbumin variants or some subgroups of immunoglobulin light chains. In other cases, incomplete degradation of the precursor has been attributed to an impaired cellular function. This has been shown with the degradation of SAA by monocytes and macrophages in vitro.[39] Several amyloid-related substances of diverse chemical nature may be involved in the processes of amyloid formation and deposition. These include the glycoprotein P component,[40,41] glycosaminoglycans,[42]

inorganic components,[43] and other substances not yet fully characterized as the amyloid enhancing factor (AEF) in experimentally-induced amyloidosis in rodents.[44,45]

The most common type of amyloidosis is Alzheimer's Disease (AD) accounting for more than 50% of the cases of dementia and affecting more than 5% of the population above the age of 65 years.[46] Moreover, amyloid deposition in the brain is present in several neurologic diseases of diverse etiology and is a concomitant finding of normal aging.[47-49] Fibrillar proteins in AD are distributed in three major lesions: 1) neurofibrillary tangles (NFT); 2) senile plaques; and 3) cerebrovascular amyloid.

NFT are intracellular amyloid deposits found in the cytoplasm of pyramidal neurons in the cerebral cortex, hippocampus and nuclei of the basal forebrain.[50] At the electron microscopy level, NFT are composed of bundles of twisted fibrils called paired helical filaments (PHF).[51,52] PHF usually show a periodic crossing over of 60 to 100 nm, a maximum width of 20 nm and a minimum width of 10 nm. Ultrastructural studies have shown that each strand of PHF appears to be composed of 3 nm subunits axially assembled in a helical fashion.[53,54] The typical senile plaque is a spherical lesion found in the cerebral cortex presenting three major constituents: 1) a central amyloid core; 2) degenerated neuronal processes; and 3) glial cells.[55] The last two forming a rim that surrounds the amyloid core. The abnormal neurites contain PHF similar to those found in the neuronal perykarion. Amyloid plaque cores are made up of typical "straight" 10 nm amyloid fibrils with a loose, random twisting. Studies of replicas obtained by quick-freezing of amyloid plaques from AD and Down's Syndrome (DS) have shown that these fibrils are hollow structures composed of 5 to 6 globular subunits assembled in a helix.[56] This type of structure resembles that described by Bladen et al.[57] for other types of amyloid fibrils. Vascular amyloid deposits are present in more than 80% of the patients and affect mainly small arteries of the leptomeninges and cerebral cortex.[58] The electron microscopic appearance of the fibrils is similar to that described for the senile plaque. This lesion was the first to provide a clue in the chemical nature of the amyloid deposits in AD and aged DS patients.[59,60] Biochemical and immunological studies have shown that cerebrovascular amyloid and amyloid plaque cores share the same subunit with a molecular weight of 4.5 kDa (β-protein).[59-62] Recently, the complementary DNA clone coding for β-protein has been sequenced showing a predicted precursor of 695 residues that resembles glycosylated cell surface receptors with β-protein starting at position 597.[63,64]

Amino acid composition of isolated NFT is similar to that found for plaque cores and cerebrovascular amyloid.[65-67] Moreover, Masters et al.[65] have reported isolation from fractions enriched in NFT of the same β-protein with a highly ragged amino terminal. However, these findings should be interpreted with caution since two major obstacles remain to be resolved: 1) although PHF are solubilized in 100% formic acid,[65] they are insoluble in the buffers used for further characterization by gel electrophoresis and chromatography; and 2) variable degrees of cross contamination between NFT and amyloid cores are found with the methods available for purification.[66-68] Therefore, information about the nature of PHF is mainly derived from immunological cross-reactivities. PHF have been shown to cross-react with normal constituents of the neuronal cytoskeleton. These include the 200 kDa component of neurofilaments,[69] microtubule-associated proteins (MAP) and MAP-tau polypeptides,[70-72] particularly, with phosphorylated epitopes of these proteins. In contrast, most of the antibodies raised against β-protein or homologous synthetic peptides have failed to label NFT in immunohistochemical studies.[61,73,74] Lack of cross-reactivity and morphological differences between PHF and plaque-vascular amyloid fibrils, are the major arguments against the concept that the three lesions are made up of the same amyloid subunit.

In vitro experiments on amyloid formation were initially described 15 years ago by Glenner et al.[10] using pepsin digestion of Bence Jones proteins. Proteolysis of immunoglobulin light chains, released fragments that formed fibrils with the characteristics of amyloid. Some, but not all, of the immunoglobulin light chains studied were able to form amyloid fibrils after proteolysis, suggesting that a specific primary structure was essential in determining "amyloidogenicity."[9,10,75] Since then, diverse proteins have been reported to adopt a β-pleated sheet secondary structure and polymerize as fibrils in vitro. These include insulin, glucagon, β_2-microglobulin and the neurotransmitter substance P.[76-78] With exception of the last, diverse conditions and treatments appear necessary: proteolysis in the case of immunoglobulin light chains,[10] acidification and heating for insulin and glucagon[76] and a progressive decrease in salt concentration for β_2-microglobulin.[77] The only factor that appears to be important in the case of substance P is an adequate concentration of this peptide.[78] Our studies using two synthetic peptides corresponding to the reported 28-residue sequence of Alzheimer β-protein (SP28) and to residues 12-28 (SP17) revealed that these peptides formed Congo red positive fibrils under physiological conditions, using different salt concentrations and in aqueous solutions.[74] No further treatment was necessary. The assembly was faster than that reported for other proteins and the time of incubation did not affect the morphology. On the contrary, it has been shown that immunoglobulin light chains present different stages of fibrillar development depending on the time of incubation.[75] Under electron microscopy SP28-17 fibrils closely resembled native amyloid fibrils isolated from plaque cores and cerebrovascular amyloid, not only in the individual fibril morphology, but also in the way these fibrils were arranged forming the typical bundles that are seen in the negatively stained extracted fractions from AD brains (Figs. 1 & 2).

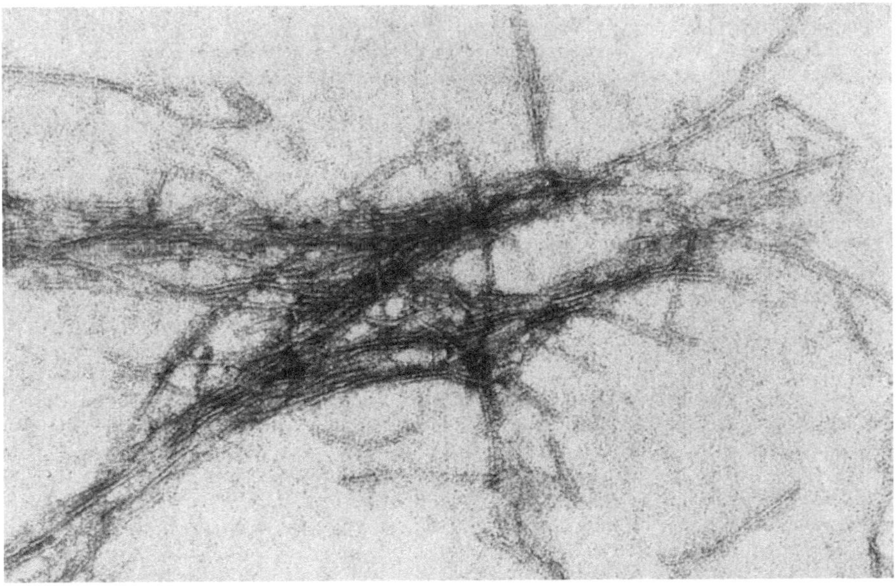

Fig. 1. Electron micrograph showing in vitro formed amyloid fibrils from SP17. Magnification x 170,000.

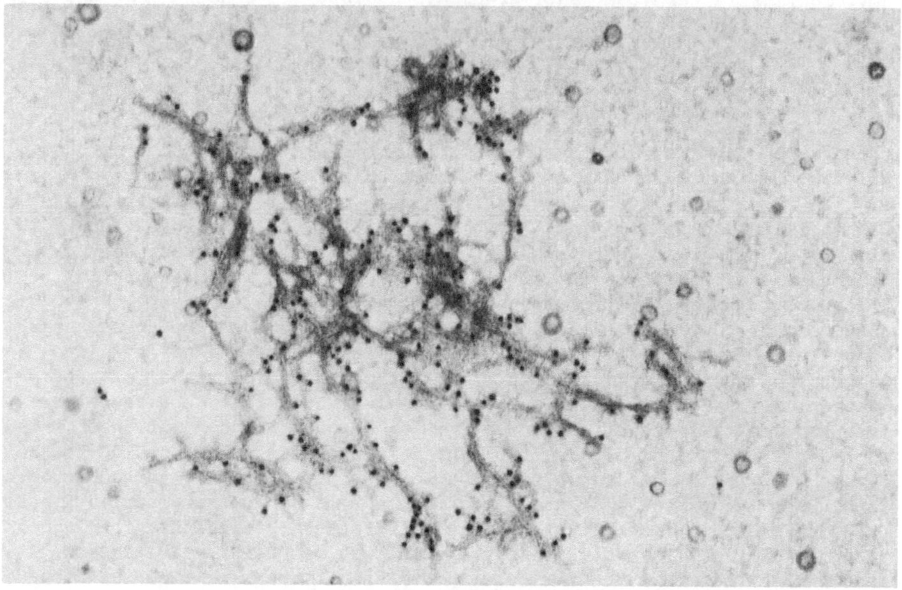

Fig. 2. Electron micrograph. Immunogold labelling of SP28 amyloid-like fibrils with anti-SP28. Magnification x 85,000.

As expected, X-ray diffraction pattern of SP28-17 was similar to that described for amyloid plaques and NFT[79] with typical reflections of β-pleated sheet secondary structure. We have also observed that a synthetic peptide of 14 residues including positions 15-28 of β-protein still formed Congo red positive birefringent fibrils with similar electron microscopic morphology. This finding is in agreement with the reported fibril formation from a peptide including residues 10-20 of β-protein.[80] In spite of the morphological similarity between synthetic and native amyloid fibrils, a marked difference in the solubility pattern was observed: synthetic fibrils were completely solubilized in SDS-containing buffers, assessed by gel electrophoresis, while relative insolubility in dissociating agents is a well known feature of AD amyloid fibrils. Several factors may be postulated to explain this finding: 1) a highly hydrophobic carboxy terminal of 14 residues (insolubility fragment) is present in the native β-protein;[63] 2) tissue-related factors may interact with the β-protein such as inorganic components (aluminum, silicon, calcium) that are known to be present in senile plaques[43] or glycosaminoglycans;[42] and 3) alterations of the β-protein due to the extraction procedure, although this seems unlikely in light of its biological behavior in vivo.

These experiments suggest that folding and self-aggregation of a small portion of the β-protein included in positions 15-28, leads to the formation of a quaternary fibrillar structure, whereas the high insolubility of the molecule is related to a different portion of the sequence that constitutes a hydrophobic tail of 14 residues (positions 29-42)(Fig. 3).

At the present time, a widely held argument against the notion of β-protein forming the three lesions in AD is the striking morphological differences between PHF and other types of amyloid. In vitro studies have shown that the same protein can adopt different fibrillar configurations depending on the environmental conditions, the degree of heterogeneity of the primary structure and diverse enzymatic treatments. This has been shown with purified MAP-tau protein adopting different morphology upon

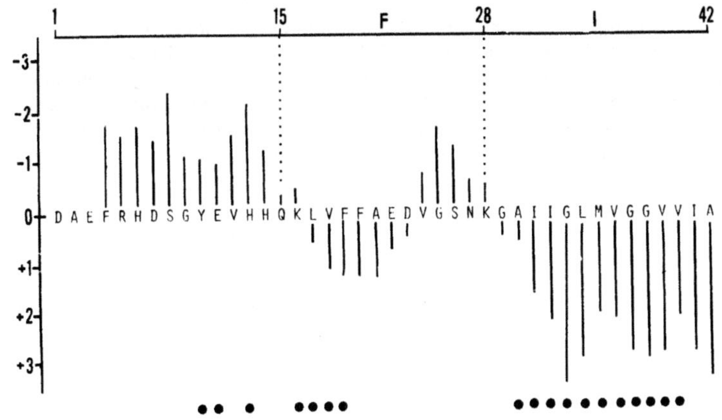

Fig. 3. Hydrophatic analysis of β-protein or A4.[84]
F: A portion included in residues 15-28 is necessary for fibril formation; I: The carboxy terminal (positions 28-42) renders β-protein insoluble; ●: Predicted β-pleated sheet secondary structure.[85]

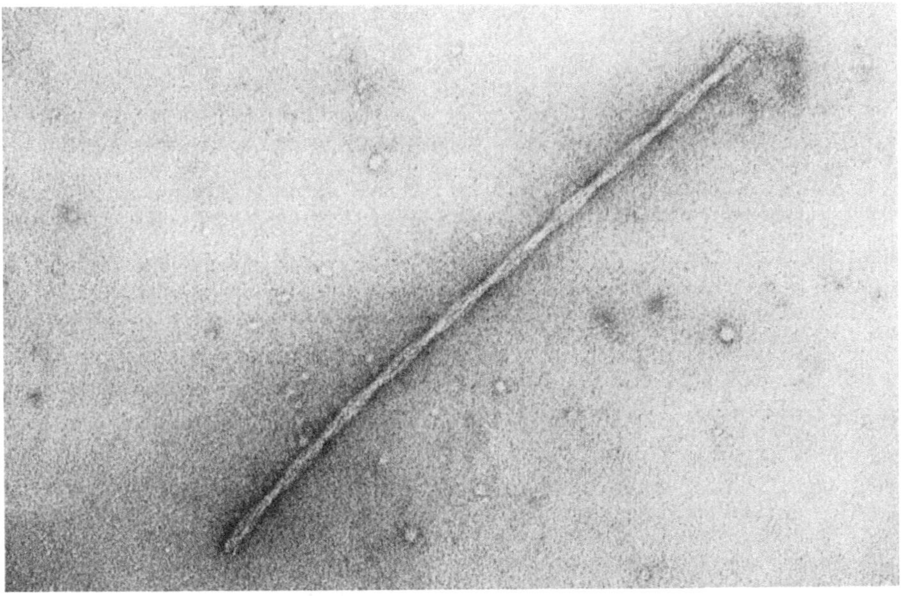

Fig. 4. Electron micrograph showing twisted amyloid AA fibrils (patient NOR) with a crossing-over of 85 nm. Magnification x 170,000.

treatment with glutaminase or after dialysis against 6M urea.[81] In addition, we have observed that amyloid protein AA isolated from the thyroid of a patient with Familial Mediterranean Fever (NOR), could reassemble to form twisted fibrils after the denaturing agents were removed by dialysis.[82] The morphology and crossing-over of AA twisted fibrils closely resembled those described for PHF (Fig. 4). Thus, the possibility remains that amino terminal heterogeneity,[65] together with unique intraneuronal conditions including cytoskeletal components, i.e., tau protein, may interact in determining the peculiar assembly of β-protein subunits to make intracellular amyloid in AD.

Alzheimer's Disease seems to be a threshold phenomenon since no pathological lesions other than those present in aged normal brains are found. Therefore, unraveling the mechanisms of amyloid formation in AD would provide not only an approach for the prevention and treatment of this major public health problem, but also an insight into the processes of normal neuronal aging.

REFERENCES

1. Glenner GG: Amyloid deposits and amyloidosis: the β-fibrilloses. N Engl J Med 302:1283-1292, 1980.
2. Glenner GG, Eanes ED, Page DL: The relation of the properties of Congo red stained amyloid fibrils to the β-conformation. J Histochem Cytochem 20:821-826, 1972.
3. Isobe T, Osserman EF: Patterns of amyloidosis and their association with plasma cell dyscrasia, monoclonal immunoglobulins and Bence Jones proteins. N Engl J Med 290:473-477, 1984.
4. Hansotia P, Ganik K, Friedenberg W: Cerebrospinal fluid monoclonal gammopathy in multiple myeloma and Waldenstrom's macroglobulinemia. Neurology (NY) 33:1411-1415, 1983.
5. Kyle RA, Greipp PR: Amyloidosis (AL) clinical and laboratory features in 229 cases. Mayo Clinic Proc 58:665-683, 1983.
6. Glenner GG, Terry W, Harada M, et al: Amyloid fibril proteins: Proof of homology with immunoglobulin light chains by sequence analysis. Science 172:1150-1151, 1971.
7. Sletten K, Husby G, Natvig JB: N-terminal amino acid sequence of amyloid fibril protein AR, prototype of a new λ-variable subgroup $V_{\lambda VI}$. Scand J Immunol 3:833-836, 1974.
8. Solomon A, Frangione B, Franklin EC: Bence Jones proteins and light chains of immunoglobulins: Preferential association of the $V_{\lambda VI}$ subgroup of human light chains with amyloidosis AL (λ). J Clin Invest 70:453-460, 1982.
9. Linke RP, Tischendorf FW, Zucker-Franklin D, et al.: The formation of amyloid-like fibrils in vitro from Bence Jones proteins of the λI subclass. J Immunol 111:24-26, 1973.
10. Glenner GG, Ein D, Eanes ED, et al: Creation of "amyloid" fibrils from Bence Jones proteins in vitro. Science 174:712-714, 1971.
11. Franklin EC, Pras M, Levin M, et al: The partial amino acid sequence of the major low molecular weight component of two human amyloid fibrils. FEBS Lett 22:121-123, 1972.
12. Pras M, Franklin EC, Shibolet S, et al: Amyloidosis associated with renal cell carcinoma is of the AA type. Am J Med 73:426-428, 1982.
13. Sohar E, Gafni J, Pras M, et al: Familial Mediterranean Fever: A survey of 470 cases and review of the literature. Am J Med 43: 227-253, 1967.
14. Benditt EP, Erikssen N, Hermodsen NA, et al: The major proteins of human and monkey amyloid substance: Common properties including unusual N-terminal amino acid sequences. FEBS Lett 19:169-173, 1971.
15. Levin M, Franklin EC, Frangione B, et al: The amino acid sequence of a major non-immunoglobulin component of some amyloid fibrils. J Clin Invest 51:2773-2776, 1972.

16. Levin M, Pras M, Franklin EC: Immunologic studies of the major non-immunoglobulin protein of amyloid. Identification and partial characterization of a related serum component. J Exp Med 138:373-380, 1973.
17. Rosenthal CJ, Franklin EC, Frangione B, et al: Isolation and partial characterization of SAA. An amyloid related protein from human serum. J Immunol 116:1415-1418, 1976.
18. Benditt EP, Erikssen N: Amyloid protein SAA is associated with high density lipoprotein from human serum. Proc Natl Acad Sci (USA) 74:4025-4028, 1977.
19. Rosenthal CJ, Franklin EC: Variation with age and diseases of an amyloid A protein-related serum component. J Clin Invest 55:746-753, 1975.
20. Sipe JD, Colten HR, Goldberger G, et al: Human serum amyloid A (SAA): Biosynthesis and postsynthetic processing of pre-SAA and structural variants defined by complementary DNA. Biochemistry 24:2931-2936, 1985.
21. Lavie G, Zucker-Franklin D, Franklin EC: Degradation of serum amyloid A protein by surface-associated enzymes of human blood monocytes. J Exp Med 148:1020-1031, 1978.
22. Sletten K, Westermark P, Natvig JB: Senile cardiac amyloid related to prealbumin. Scan J Immunol 12:503-506, 1980.
23. Husby G, Ranlow PJ, Sletten K, et al: The amyloid in familial amyloid cardiomyopathy of Danish origin is related to prealbumin. Clin Exp Immunol 60:207-216, 1985.
24. Saraiva MJM, Birken S, Costa PP, et al: Amyloid fibril protein in familial amyloidotic polyneuropathy Portuguese type. J Clin Invest 74:104-119, 1984.
25. Tawara S, Nakazato M, Kangawa K, et al: Identification of amyloid prealbumin variant in familial amyloidotic polyneuropathy (Japanese type). Biochem Biophys Res Commun 116:880-888, 1983.
26. Dwulet FE, Benson MD: Primary structure of an amyloid prealbumin and its plasma precursor in a heredofamilial polyneuropathy of Swedish origin. Proc Natl Acad Sci (USA) 81:694-698, 1984.
27. Pras M, Prelli F, Gafni J, et al: Genetic heterogeneity of familial amyloid polyneuropathies of Jewish type, in Glenner GG, Osserman EF, Benditt EP, et al (eds): Amyloidosis, Plenum Press, NY, 1986, p 385.
28. Dwulet FE, Benson MD: Characterization of a transthyretin (prealbumin) variant associated with familial amyloidotic polyneuropathy type II (Indiana/Swiss). J Clin Invest 78:880-886, 1986.
29. Grey HM, Kubo RT, Colon SM, et al: The small subunit of HL-A antigens is β_2-microglobulin. J Exp Med 138:1608-1612, 1973.
30. Poulik MD, Perry DJ, Sekine T, et al: Statistical analysis of β_2-microglobulin levels in sera of lung and gastrointestinal cancer patients. Vox Sang 38:328-333, 1980.
31. Gejyo F, Yamada T, Odani S, et al: A new form of amyloid protein with chronic hemodialysis was identified as β_2-microglobulin. Biochem Biophys Res Commun 129:701-706, 1985.
32. Gorevic PD, Munoz P, Casey TT, et al: Polymerization of intact β_2-microglobulin in tissue causes amyloidosis in patients on chronic hemodialysis. Proc Natl Acad Sci (USA) 83:7908-7912, 1986.
33. Sletten K, Westermark P, Natvig JB: Characterization of amyloid fibril proteins from medullary carcinoma of the thyroid. J Exp Med 143: 993-997, 1976.
34. Cohen DH, Feiner H, Jensson O, et al: Amyloid fibril in hereditary cerebral hemorrhage with amyloidosis (HCHWA) is related to the gastroentero-pancreatic neuroendocrine protein γ-trace. J Exp Med 158:623-628, 1983.
35. Ghiso J, Pons-Estel B, Frangione B: Hereditary cerebral amyloid angiopathy: The amyloid fibrils contain a protein which is a variant of Cystatin C, an inhibitor of lysosomal cysteine proteases. Biochem Biophys Res Commun 136:548-554, 1986.

36. Ghiso J, Jensson O, Frangione B: Amyloid fibrils in hereditary cerebral hemorrhage with amyloidosis of Icelandic type is a variant of γ-trace basic protein (Cystatin C). Proc Natl Acad Sci (USA) 83:2974-2978, 1986.
37. Wattendorff AR, Bots GThAM, Went LN: Familial cerebral amyloid angiopathy presenting as recurrent cerebral hemorrhage. J Neurol Sci 55:121-135, 1982.
38. Luyendijk W, Bots GThAM, Veger-Van Der Vlis M, et al: Familiaire hersenbloedingen als gevolg van cerebrale amyloide angiopathie. Ned Tijdschr Geneeskd 130:1935-1940, 1986.
39. Zucker-Franklin D, Fuks A: The role of the macrophage phagocytic system (MPS) in the development of secondary amyloidosis, in Marrink J, Van Rijswijk, MH (eds): Amyloidosis, Dordrecht, Martinus Nijhoff, 1986, p 329.
40. Shirahama T, Skinner M, Sipe JD, et al: Widespread occurrence of AP in amyloidotic tissues. Virchows Arch 48:197-206, 1985.
41. Prelli F, Pras M, Frangione B: The primary structure of human tissue amyloid P component from a patient with primary idiopathic amyloidosis. J Biol Chem 260:12895-12898, 1985.
42. Snow AD, Willmer J, Kisilevsky R: Sulfated glycosaminoglycans: A common constituent of all amyloids? Lab Invest 56:120-124, 1987.
43. Candy JM, Oakley AE, Klinowski J, et al: Aluminum silicates and senile plaque formation in Alzheimer's Disease. Lancet 1:354-357, 1986.
44. Axelrad MA, Kisilevsky R: Biological characterization of amyloid-enhancing factor, in Glenner GG, Costa PP, Freitas F (eds): Amyloid and Amyloidosis. Amsterdam, Excerpta Medica, 1980, p 527.
45. Hol PR, Snel FWJJ, Niewold ThA, et al: Amyloid enhancing factor (AEF) in the pathogenesis of AA-amyloidosis in the hamster. Virchows Arch 52:273-281, 1981.
46. Katzman R: The prevalence and malignancy of Alzheimer's Disease: A major killer. Arch Neurol 33:217-219, 1976.
47. Lampert PW, Gajdusek DC, Gibbs CJ Jr: Subacute spongiform virus encephalopathies: Scrapie, Kuru and Creutzfeldt-Jakob disease: A review. Am J Pathol 68:626-646, 1972.
48. Wisniewski K, Jervis GA, Moretz RC, et al: Alzheimer neurofibrillary tangles in diseases other than senile or presenile dementia. Ann Neurol 5:288-294, 1979.
49. Stam FC, Wigboldus JM, Smeulders WM: Age incidence of senile brain amyloidosis. Path Res Pract 181:558-562, 1986.
50. Selkoe DJ, Ihara Y, Abraham C, et al: Biochemical and immunocytochemical studies of Alzheimer paired helical filaments, in Katzman R (ed): Biological aspects of Alzheimer's Disease, Banbury Report 15, Cold Spring Harbor, 1983, p 125.
51. Kidd M: Paired helical filaments in electron microscopy of Alzheimer's Disease. Nature 197:192-193, 1963.
52. Wisniewski HM, Narang HK, Terry RD: Neurofibrillary tangles of paired helical filaments. J Neurol Sci 27:173-181, 1976.
53. Wisniewski HM, Wen GY: Substructures of paired helical filaments from Alzheimer's Disease neurofibrillary tangles. Acta Neuropath (Berl) 66:173-176, 1985.
54. Crowther RA, Wischik CM: Structure of the Alzheimer paired helical filament, in Marrink J, Van Rijswijk MH (eds): Amyloidosis, Dordrecht, Martinus Nijhoff, 1986, p 159.
55. Terry RD, Gonatas NK, Weiss M: Ultrastructural studies in Alzheimer's presenile dementia. Am J Pathol 44:269-297, 1964.
56. Miyakawa T, Watanabe K, Katswagi S: Ultrastructure of amyloid fibrils in Alzheimer's Disease and Down's Syndrome. Virchows Arch 52: 99-106, 1986.
57. Bladen HA, Nylen MU, Glenner GG: The ultrastructure of human amyloid as revealed by the negative staining technique. J Ultrastruct Res 14:449-459, 1966.

58. Mandybur TI: The incidence of cerebral amyloid angiopathy in Alzheimer's Disease. Neurology 25:120-126, 1975.
59. Glenner GG, Wong CW: Alzheimer's Disease: Initial report of the purification and characterization of a novel cerebrovascular amyloid protein. Biochem Biophys Res Commun 120:885-890, 1984.
60. Glenner GG, Wong, CW: Alzheimer's Disease and Down's Syndrome: sharing of a unique cerebrovascular amyloid fibril protein. Biochem Biophys Res Commun 122:1131-1135, 1984.
61. Wong CW, Quaranta V, Glenner GG: Neuritic plaques and cerebrovascular amyloid in Alzheimer's Disease are antigenically related. Proc Natl Acad Sci (USA) 82:8729-8732, 1985.
62. Masters CL, Simms G, Weinman NA, et al: Amyloid plaque core protein in Alzheimer's Disease and Down's Syndrome. Proc Natl Acad Sci (USA) 82:4245-4249, 1985.
63. Kang J, Lemaire HG, Unterbeck A, et al: The precursor of Alzheimer's Disease amyloid A4 protein resembles a cell-surface receptor. Nature 325:733-736, 1987.
64. Goldgaber D, Lerman MI, McBride OW, et al: Characterization and chromosomal localization of a cDNA encoding brain amyloid of Alzheimer's Disease. Science 235:877-880, 1987.
65. Masters CL, Multhaup G, Simms G, et al: Neuronal origin of a cerebral amyloid: Neurofibrillary tangles of Alzheimer's Disease contain the same protein as the amyloid of plaque core and blood vessels. EMBO J 4:2757-2763, 1985.
66. Gorevic PD, Goni F, Pons-Estel B, et al: Isolation and partial characterization of neurofibrillary tangles and amyloid plaque core in Alzheimer's Disease - immunohistological studies. J Neuropath Exp Neurol 45:647-664, 1986.
67. Allsop D: Biochemistry of cerebral amyloid in Alzheimer's Diease, the unconventional slow virus diseases and Icelandic cerebrovascular amyloidosis, Marrink J, Van Rijswijk MH (eds): Amyloidosis, Dordrecht, Martinus Nijhoff, 1986, p 243.
68. Selkoe DJ, Ihara Y, Salazar FJ: Alzheimer's Disease: Insolubility of partially purified paired helical filaments in sodium dodecyl sulfate and urea. Science 215:1243-1245, 1982.
69. Perry G, Rizzuto N, Autilio-Gambetti L, et al: Paired helical filaments from Alzheimer's Disease patients contain cytoskeletal components. Proc Natl Acad Sci (USA) 82:3916-3920, 1985.
70. Grundke-Iqbal I, Iqbal K, Quinlan M, et al: Microtubule associated protein tau: A component of Alzheimer paired helical filaments. J Biol Chem 261:6084-6089, 1986.
71. Wood JG, Mirra SS, Pollock NJ, et al: Neurofibrillary tangles of Alzheimer's Disease share antigenic determinants with the axonal microtubule-associated protein tau (t). Proc Natl Acad Sci (USA) 83:4040-4043, 1986.
72. Kosik KS, Joachin CL, Selkoe DJ: Microtubule associated protein tau (t) is a major antigenic component of paired helical filaments in Alzheimer's Disease. Proc Natl Acad Sci (USA) 83:4044-4048, 1986.
73. Allsop D, Landon M, Kidd M, et al: Monoclonal antibodies raised against a subsequence of senile plaque core protein react with plaque cores, plaque periphery and cerebrovascular amyloid in Alzheimer's Disease. Neurosc Lett 68:252-256, 1986.
74. Castano EM, Ghiso J, Prelli F, et al: In vitro formation of amyloid fibrils from two synthetic peptides of different lengths homologous to Alzheimer's Disease β-protein. Biochem Biophys Res Commun 141: 782-789, 1986.
75. Shirahama T, Benson MD, Cohen AS, et al: Fibrillar assemblage of variable segments of immunoglobulin light chains: An electron microscopic study. J Immunol 110:21-30, 1973.
76. Glenner GG, Eanes ED, Bladen HA, et al: β-pleated sheet fibrils: a comparison of native amyloid with synthetic protein fibrils. J Histochem Cytochem 22:1141-1158, 1974.

77. Connors LH, Shirahama T, Skinner M, et al: In vitro formation of amyloid fibrils from intact β_2-microglobulin. Biochem Biophys Res Commun 131:1063-1068, 1985.
78. Perry EK, Oakley AE, Candy JM, et al: Properties and possible significance of substance P and insulin fibrils. Neurosc Lett 25:321-325, 1981.
79. Kirschner CA, Abraham C, Selkoe DJ: X-ray diffraction from intraneuronal paired helical filaments and extraneuronal amyloid fibers in Alzheimer's Disease indicates cross-B conformation. Proc Natl Acad Sci (USA) 83:503-507, 1986.
80. Masters CL, Beyreuther K: Amyloidogenic proteins in human central nervous system diseases, in Marrink J, Van Rijswijk MH (eds): Amyloidosis, Dordrecht, Martinus Nijhoff, 1986, p 149.
81. Garcini EM, Serrano L, Avila J: Self assembly of microtubule associated protein tau into filaments resembling those found in Alzheimer's Disease. Biochem Biophys Res Commun 141:790-796, 1986.
82. Prelli F, Pras M, Frangione B: The degradation and deposition of amyloid AA fibril is tissue specific. Submitted for publication, 1987.
83. Bendheim PE, Bockman JM, McKinley MP: Scrapie and Creutzfeldt-Jakob disease prion proteins share physical properties and antigenic determinants. Proc Natl Acad Sci (USA) 82:997-1001, 1985.
84. Kyle P, Doolittle RF: A simple method for displaying the hydropathic character of a protein. J Mol Biol 157:105-132, 1982.
85. Chou PY, Fasman GD: Prediction of protein conformation. Biochemistry 13:222-245, 1974.

ACKNOWLEDGEMENTS

This work was supported by United States Public Health Service Grants AG 05891, AM 02594 and AM 01431.

CLONING OF NEUROFIBRILLARY TANGLE-RELATED GENES

Kenneth S. Kosik and Rachael Neve

Center for Neurologic Diseases, Department of Medicine
Brigham and Women's Hospital
Neurosciences Department, Children's Hospital Medical Center
Harvard Medical School

INTRODUCTION

In the past few years it has become increasingly clear that the neuronal microtubular system is involved in the pathogenesis of Alzheimer neurofibrillary tangles (NFT). Electron microscopists were among the first to observe the relative paucity of microtubules in NFT-containing neurons. The microtubular system was later implicated immunochemically, based on an antiserum that cross-reacted with NFT and microtubules.[1] In 1980 Eng et al.[2] demonstrated that the cross-reactive element between microtubules and NFT was not the tubulin, which is the most abundant component of microtubules, because antibodies to tubulin did not cross-react with NFT. Attention turned to the microtubule-associated proteins (MAPs) which are a class of proteins that cycle with microtubules through temperature-dependent phases of polymerization/depolymerization and are morphologically associated with the microtubule system. One of the possible functions of the MAPs is to stabilize the microtubules against depolymerization[3] and modulate their dynamic instability.[4] Relative to other tissues the brain contains an abundance of microtubules and appears to have evolved its own system of MAPs. Brain MAPs have been divided into a high molecular weight group and a heterogeneous protein of 55-62 kDa designated tau. As seen on SDS-PAGE the high molecular weight group consists of a complex series of bands designated MAP1[5] and a doublet of approximately 280 kDa designated MAP2. Additional MAPs with unique tissue distributions and developmental chronologies have been described as well as "light chains" and kinases that bear a close association with those MAPs mentioned above.[6] Of the various MAPs in brain, MAP2 is the most abundant.

MAPs were shown to be involved in certain experimentally-induced filamentous accumulations. Under these conditions MAPs lose their association with tubulin in favor of collecting within filamentous aggregates. This phenomenon was demonstrated in vinblastine-treated axons.[8-12] In the former case MAP-immunoreactivity partitioned with aggregates of intermediate filaments collapsed into perinuclear bundles; in the later case MAP-immunoreactivity segregated with the neurofilaments to the periphery of the axon while intact microtubules capable of supporting fast transport remained intact in the central portion of the axon.

In 1984 we demonstrated that NFT selectively incorporate certain MAP2 epitopes.[13] While these epitopes were present within the mesh of aggregated filamentous structures which comprise the NFT (an area devoid of microtubules), MAP2-immunoreactivity was not present when NFT were extracted with the ionic detergent, sodium dodecyl sulfate. This finding suggested that MAP2 might not be associated with the insoluble ultrastructural component of NFT, the PHF. Subsequently several laboratories independently discovered that the tau protein is an integral component of the PHF.[14-19] This suggestion was based on the finding that NFT on tissue sections as well as SDS-extracted PHF were tau-immunoreactive and that when SDS-extracted PHF were injected into rabbits a tau-reactive anti-serum was consistently obtained. Anti-serum from some rabbits injected with SDS-extracted material displayed immunoreactivity to both tau and MAP2. Several lines of evidence suggest that there is more to PHF than an insoluble form of polymerized tau: Tau antibodies do not decorate ghost tangles very well, PHF antiserum absorbed with tau protein (and MAP2) continues to immunoreact with NFT, and SDS-extracted PHF treated with trypsin lose their tau-immunoreactivity while maintaining their ultrastructural integrity.[20] This latter finding raises the possibility that tau may not be structurally essential to the architecture of the PHF. Neurofilaments have also been observed to share epitopes with NFT.[21-25] An integral role for the neurofilament as a component of the PHF was challenged by the finding that the subset of neurofilament monoclonal antibodies which cross-react with NFT, also cross-react with tau.[26] This subset of neurofilament antibodies is directed against the phosphorylated form of the protein and as such cross-reacted with the phosphorylated epitope(s) of tau protein. One such antibody from this class of phosphoneurofilament antibodies also cross-reacts with the phosphoprotein, rhodopsin.[27]

Within the neuron MAP2 and tau occupy complementary distributions; MAP2 is confined to the somatodendritic compartment[28-29] and tau to the axonal compartment.[30-32] This topographical segregation is strictly observed in all neuronal cell types so far studied. During cortical development when MAP2 is first detected in the rodent brain at E14, we have noted that it is positioned within the somatodendritic compartment.[33] In Alzheimer's disease the strict topographical segregation is violated.[34] The tau-immunoreactive NFT in Alzheimer's disease cortex are generally located within the pyramidal cell soma and apical dendrite, a MAP-rich region that is normally devoid of tau-immunoreactivity.[16,34] Most remarkably revealed by tau immunocytochemistry of Alzheimer cerebral cortex is the extensive and dense neuritic pathology which appears more abundant than either the senile plaques or the NFT. Numerous swollen dystrophic neurites are present throughout the neuropil that react intensely with our tau monoclonal antibody.[34] These abnormal neurites are present in all cortical layers even in regions without amyloid deposition. They extend beyond the regions of the senile plaque as seen by their density in the molecular layer which generally has a paucity of senile plaques. Whether these tau-immunoreactive neurites (unassociated with senile plaques) are highly distorted and retracted axons or dendrites or both is as yet unanswered. The former case suggests a complete structural disintegration of the axon, the latter an aberrantly transported tau to the distal extent of the dendrite. Compared to control cases, one can appreciate the extent to which the normal axonal architecture is distorted. In the Alzheimer brain the radial deployment of axons is lost with relatively few normal appearing intra-cortical axonal fibers. When one visualizes the complementary side of the neuronal architecture by staining the somatodendritic compartment with a MAP2 monoclonal antibody, there is a relative preservation of the vertically ascending pyramidal cell dendritic bundles.[35]

The expression and regulation of the amyloid beta-protein gene on chromosome 21 has been posited as fundamental to the pathogenesis of Alzheimer's disease, based upon linkage analysis of several large Alzheimer kindreds.[36-40] Senile plaques containing amyloid that is antigenically identical to Alzheimer amyloid are abundantly present in the brains of aged non-human primates.[41] These species do not, however, develop paired helical filaments (PHF). Therefore, to the extent that the NFT and dystrophic neurites contribute to dementia their unique presence in the human in association with the amyloid needs to be explained. To gain the structural information required to understand the molecular construction of the neurofibrillary tangle we have cloned the genes of some of the proteins thought to be involved and have begun to sequence their cDNA's.

CLONING OF THE HUMAN MAP2 GENE

Monoclonal antibodies, 5F9 and 4F7, were described previously[13] as specifically directed against distinct epitopes within MAP2. While 5F9 reacted with Alzheimer NFT, 4F7 did not. The antibodies were pooled to screen a human fetal brain expression library in λgt11. A 22-week male abortus obtained under approved protocols was used to prepare RNA. The resultant library, amplified in E. coli Y1088 contained 7.5 million independent recombinant phage. cDNA was synthesized from 10 μg of polyadenylated RNA by a modification of the method of Gubler and Hoffman[42] described in reference 43. To screen the library, the phage were grown on Y1090 at a density of 100,000 per 150 mm^2 plate and standard protocol followed for incubation of the filters in primary antibody.[44-46] The filters were incubated in peroxidase-conjugated goat anti-mouse IgG (1:500) and visualized with diaminobenzidine and hydrogen peroxide. A clone designated KN7 reacted with both antibodies individually and was plaque purified. The insert DNA was 2.5 kilobases (kb). On Southern blots[47] in which KN7 was hybridized to EcoRI digests of total human DNA bands at 7.4 kb and 1.2 kb were detected. The identity of the clone was confirmed by matching the direct amino acid sequence of a MAP2 trypsin fragment with a portion of the deduced amino acid sequence from the cloned cDNA. The fusion protein produced by this cDNA consisted of a 93 kDa insert bound to beta-galactosidase. The fusion protein was also recognized by a third MAP2 antibody, which is a polyclonal antibody raised against highly purified MAP2.

To study the tissue localization of MAP2 expression Northern analyses were performed. KN7 was hybridized to a panel of total RNA extracted from 10 fetal tissues. Hybridization was detected in a single 9 kb band present only in fetal brain. The question as to whether the MAP2 message was confined to neurons in brain was studied with RNA prepared from neuronal cultures. Neuronal cultures from rat cortex were either left untreated or treated with kainic acid to reduce or eliminate neuronal populations. Total RNA was extracted from the untreated and treated neuronal cultures as well as from astrocyte and fibroblast cultures. The MAP2 cDNA hybridized only to the untreated neuronal cultures. A comparison of the expression of the MAP2 message between adult and embryonic rat brain was analyzed by RNA blots and compared with MAP2 protein expression as detected by immunoblots of SDS-gels. KN7 hybridized to a 9 kb band in both adult and embryonic rat brain; however a more intense hybridization occurred in the adult. This pattern was paralleled on the Western blots which indicated only trace MAP2 at approximately 20 weeks gestation. Human MAP2 at early developmental time points is a single band, similar to the electrophoretic pattern seen in rat and cow. An immunocytochemical description of the distribution of MAP2 in human fetal cerebral cortex has recently been reported by Sims et al.[48]

The cDNA was used for chromosomal localization by both spot blots of sorted chromosomes and by in situ hybridization.[49] Spot blots were done by Dr. Peter Harris who hybridized KN7 to a panel of 10 nitrocellulose filters, onto each of which two human chromosome fractions had been separately sorted. Specific hybridization was seen only to the chromosome 2 fraction. To localize precisely the MAP2 gene on chromosome 2, in situ hybridization was done by Dr. Timothy Donlon using methods described by Harper and Saunders[50] and modified by Donlon et al.[51] The isolated cDNA, at a concentration of 1 pg/µl was radiolabeled by the random priming method.[52] One hundred metaphase spreads were examined: 40% displayed silver grains over the long arm of chromosome 2 at bands 2q34-35 and 41 of 216 grains (19%) were localized to this same region.

CLONING OF THE HUMAN TAU GENE

A mouse tau cDNA denoted pTA2 was generously provided to us by G. Lee, D. Drubin and M. Kirschner for hybridization to the human fetal brain library described above. Phage were plated on the strain Y1088 at a density of 10^5 per 150 mm plate. The insert DNA was released with PstI and labeled with alpha [^{32}P]-dCTP by random oligonucleotide priming[52] to a specific activity greater than 109 cpm/µg. Plaque lifts and hybridization were done according to the protocol of Benton and Davis.[52] The hybridization resulted in 75 putative human tau cDNA clones under stringent conditions (65°C, 4x SSC, and no formamide). Among the seven chosen for further analysis the insert size ranged from 1.0-2.9 kb in length as determined by EcoRI digestion of the cDNAs and electrophoresis on 0.8% agarose gels. To confirm the identity of these clones two of them designated p9TAU (2.9 kb) and p16TAU (1.7 kb) were subcloned into the plasmid, pBR322, and hybridized to identical Southern blots[47] of mouse DNA cleaved with five different restriction enzymes. The two human clones and pTA2 all hybridized to a 6.5 kb EcoRI band. p9TAU showed more extensive homology with pTA2 than p16TAU since it also shared a 10 kb BamHI band, a 3.5 kb BamHI band, a 7 kb HindIII band, a 2.2 kb HindIII band, a 5.5 kb Pst band, a 4 kb TaqI band, and a 2.5 TaqI band. p9TAU and p16TAU cross-hybridize with each other and restriction analysis of these cDNAs reveals a common region of approximately 500 base pairs. p9TAU was definitively identified as a human tau by dideoxy sequencing a fragment subcloned into M13 and demonstrating its very high homology with a region within the mouse tau sequence as shown to us by Dr. Gloria Lee. Direct protein sequence from purified trypsin fragments of bovine tau have also served to confirm the identity of the clone.

The distribution of the tau message was studied by Northern blot analysis. Both tau cDNAs hybridized to a single 6 kb human fetal brain mRNA and did not hybridize to any other total RNA extracted from 10 different fetal tissues. A developmental increase in the amount of tau message was shown in rat to which the probes cross-hybridized, by comparing embryonic day 16 rat with mature rat; equally loaded Northern blots revealed markedly increased signal in the mature animal. A comparison of the message size in fetal and adult rat indicated that the adult mRNA band was approximately 6.5 kb and the fetal band was approximately 6.0 kb. This data suggests that the fetal to adult transformation of tau occurs, at least in part, at the level of the messenger RNA and cannot solely be attributed to a post-translational modification. This large message for the tau is similar to that described in mouse for which a 6 kb message was also described.[54] p9TAU was hybridized to a similar set of filters as described above for spot blots to determine the chromosomal localization of tau. Specific hybridization to chromosome 17 was observed. When p16TAU was used as the probe chromosome 6 and 17 spots gave positive signals. The site of the human tau gene was also determined by in situ hybridization as described above for the MAP2 cDNA. p9TAU mapped to the long arm of chromosome 17 (15 of 50 cells contained grains in this region) and 17 of 104 grains (16.3%) were observed at band 17q21. As suggested by the spot blots, p16TAU hybridized to 17q21 and 6p21. The additional localization of p16TAU to the short arm of chromosome 6 raises the possibility of more than one tau gene or homology of p16TAU to a non-expressed pseudo-gene (55).

REFERENCES

1. Grundke-Iqbal I, Johnson AB, Wisniewski HM, Terry RD, Iqbal K: Evidence that Alzheimer neurofibrillary tangles originate from neurotubules. Lancet 1:578-580, 1979
2. Eng LF, Forno LS, Bigbee JW, Forno KI: Immunocytochemical localization of glial fibrillary acidic protein and tubulin in Alzheimer's disease brain biopsy tissue, in Amaducci L, Davison AN, Antuono P (eds): Aging, vol 13. New York, Raven Press, 1980, pp. 49-54
3. Drubin DG, Kirschner MW: Tau protein function in living cells. J Cell Biol 103:2739-2746, 1986
4. Horio T, Hotani H: Visualization of the dynamic instability of individual microtubules by dark-field microscopy. Nature 321:605-607, 1986
5. Bloom GS, Schoenfeld TA, Vallee RB: Widespread distribution of the major polypeptide component of MAP1 (Microtubule-associated protein 1) in the nervous system. J Cell Biol 98:320-330, 1984
6. Vallee RB, Bloom GS: High molecular weight microtubule associated proteins (MAPs), in Satir B (ed): Modern Cell Biology, vol 3. New York, Alan Liss, 1984, pp. 21-75
7. Bloom GS, Vallee RB: Association of microtubule-associated protein 2 (MAP2) with microtubules and intermediate filaments in cultured brain cells. J Cell Biol 96:1523-1531, 1983
8. Papasozomenos SCh, Yoon M, Crane R, Autilio-Gambetti L, Gambetti P: Redistribution of proteins of fast axonal transport following administration of β,β'-iminodipropionitrile: A quantitative autoradiographic study. J Cell Biol 95:672-675, 1982
9. Papasozomenos SCh, Binder LI, Bender PK, Payne MR: Microtubule-associated protein 2 within axons of apical motor neurons: Associations with microtubules and neurofilaments in normal and β,β'-iminodipropionitrile-treated axons. Cell Biol 100:74-85, 1985
10. Hirokawa N, Bloom GS, Vallee RB: Cytoskeletal architecture and immunocytochemical localization of microtubule-associated proteins in regions of axons associated with rapid axonal transport: the beta, beta'-iminodipropionitrile-intoxicated axon as a model system. J Cell Biol 101:227-239, 1985
11. Griffin JW, Fahnstock KE, Price DL, Hoffman PN: Microtubule-neurofilament segregation produced by β,β'-iminodipropionitrile: Evidence for the association of fast axonal transport with microtubules. J Neurosci 3:557-566, 1983
12. Papasozomenos SC, Autilio-Gambetti L, Gambetti P: Reorganization of axoplasmic organelles following β,β'-iminodipropionitrile administration. J Cell Biol 91:866-871, 1981
13. Kosik KS, Duffy LK, Dowling MM, Abraham C, McCluskey A, Selkoe DJ: Microtubule-associated protein 2: Monoclonal antibodies demonstrate the selective incorporation of certain epitopes into Alzheimer neurofibrillary tangles. Proc Natl Acad Sci USA 81:7941-7945, 1984
14. Brion JP, Couck AM, Passareiro E, Flament-Durand J, in Traper J, Gispens WH (eds): Advances in Applied Neurological Science: Senile Dementia of the Alzheimer Type. Berlin, Springer, 1985, pp. 164-174

15. Wood JG, Mirra SS, Pollock NJ, Binder LI: Neurofibrillary tangles of Alzheimer disease share antigenic determinants with the axonal microtubule-associated protein tau (τ). Proc Natl Acad Sci USA 83: 4040-4043, 1986
16. Kosik KS, Joachim CL, Selkoe DJ: Microtubule-associated protein tau (τ) is a major antigenic component of paired helical filaments in Alzheimer disease. Proc Natl Acad Sci USA 83:4044-4048, 1986
17. Grundke-Iqbal I, Iqbal K, Quinlan M, Tung Y-C, Zaidi MS, Wisniewski HM: Microtubule-associated protein tau. J Biol Chem 216:6084-6089, 1986
18. Nukina N, Ihara Y. One of the antigenic determinants of paired helical filaments is related to tau protein. J Biochem 99:1541-1544, 1986
19. Yen S-H, Dickson DW, Crowe A, Butler M, Shelanski ML: Alzheimer's neurofibrillary tangles contain unique epitopes and epitopes in common with the heat-stable microtubule associated proteins, tau and MAP2. Amer J Pathol 126(1):81-91, 1987
20. Merz P et al.: Abstract submitted to American Association of Neuropathologists for Seattle 1987 meeting
21. Anderton BH, Breinburg D, Downes MJ, Green PJ, Tomlinson BE, Ulrich J, Wood JN, Kahn J: Monoclonal antibodies show that neurofibrillary tangles and neurofilaments share antigenic determinants. Nature 298:84-86, 1982
22. Autilio-Gambetti L, Gambetti P, Crane RC: Paired helical filaments: Relatedness to neurofilaments shown by silver staining and reactivity with monoclonal antibodies, in Katzman R (ed): Banbury Report 15: Biological aspects of Alzheimer's disease. Cold Spring Harbor, NY: Cold Spring Harbor Laboratory, 1983, pp. 117-124
23. Ihara Y, Nukina N, Sugita H, Toyokura Y: Staining of Alzheimer's neurofibrillary tangles with antiserum against 200K component of neurofilament. Proc Jap Acad 57:152-156, 1981
24. Rasool CG, Abraham C, Anderton BH, Haugh M, Kahn J, Selkoe DJ: Alzheimer's disease: Immunoreactivity of neurofibrillary tangles with anti-neurofilament and anti-paired helical filament antibodies. Brain Res 310:249-260, 1984
25. Dahl D, Selkoe DJ, Pero RT, Bignami A: Immunostaining of neurofibrillary tangles in Alzheimer's senile dementia with a neurofilament antiserum. J Neurosci 2:113-119, 1982
26. Nukina N, Kosik KS, Selkoe DJ: Recognition of Alzheimer paired helical filaments by monoclonal neurofilament antibodies is due to crossreaction with tau protein. Proc Natl Acad Sci USA (in press)
27. Balkema GW, Drager UC: Light-dependent antibody labelling of photoreceptors. Nature 316:630-633, 1985
28. Matus A, Bernhardt R, Hugh-Jones T: High molecular weight microtubule-associated proteins are preferentially associated with dendritic microtubules in brain. Proc Natl Acad Sci USA 78:3010-3014, 1981
29. Miller P, Walter U, Theurkauf WE, Vallee RB, DeCamilli P: Frozen tissue sections as an experimental system to reveal specific binding sites for the regulatory subunit of type II cAMP-dependent protein kinase in neurons. Proc Natl Acad Sci USA 79:5562-5566, 1982
30. Tytell M, Brady ST, Lasek RJ: Axonal transport of a subclass of tau proteins. Evidence for the regional differentiation of microtubules in neurons. Proc Natl Acad Sci USA 81:1570-1574, 1984
31. Binder LI, Frankfurter A, Rebhun LI: The distribution of tau in the mammalian central nervous system. J Cell Biol 101:1371-1378, 1985
32. Peng I, Binder LI, Black MM: Biochemical and immunological analyses of cytoskeletal domains of neurons. J Cell Biol 102:252-262, 1986
33. Crandall JE, Jacobsen M, Kosik KS: Ontogenesis of microtubule-associated protein 2 (MAP 2) in embryonic mouse cortex. Dev Brain Res 28:127-133, 1986

34. Kowall N, Kosik KS: The cytoskeletal pathology of Alzheimer's disease is characterized by abnormal tau distribution (submitted)
35. Kosik KS, Rogers J, Kowall NW: Senile plaques are located between apical dendritic clusters. J Neuropathol Exp Neurol 46:1-11, 1987
36. St. George-Hyslop PH, Tanzi RE, Polinsky RJ, Haines JL, Nee L, Watkins PC, Myers RH, Feldman RG, Pollen D, Drachman D, Growdon J, Bruni A, Foncin J-F, Salmon D, Frommelt P, Amaducci L, Sorbi S, Placentini S, Stewart GD, Hobbs WJ, Conneally PM, Gusella JF: The genetic defect causing familial Alzheimer's disease maps on chromosome 21. Science 235:885-890, 1987
37. Tanzi RE, Gusella JF, Watkins PC, Bruns GAP, St. George-Hyslop P, Van Keuren ML, Patterson D, Pagan S, Kurnitt DM, Neve RL: Amyloid β protein gene. cDNA, mRNA distribution and genetic linkage near the Alzheimer locus. Science 235:880-884, 1987
38. Goldgaber D, Lerman MI, McBride OW, Saffiotti U, Gajdusek DC: Characterization and chromosomal localization of a cDNA encoding brain amyloid of Alzheimer's disease. Science 235:877-879, 1987
39. Robakis NK, Wisniewski HM, Jenkins EC, Devine-Gage E, Houck GE, Yao X-L, Ramakirshna N, Wolfe G, Silverman WP, Brown WT: Chromosome 21q21 sublocalization of gene encoding beta-amyloid peptide in cerebral vessels and neuritic (senile) plaques of people with Alzheimer disease and Down syndrome. Lancet 1:385-386, 1987
40. Kang J, Lemaire H-G, Unterbeck A, Salbaum JM, Masters CL, Grzeschik K-H, Multhaup G, Beyreuther K, Müller-Hill B: The precursor of Alzheimer's disease amyloid A4 protein resembles a cell-surface receptor. Nature 325:733-736, 1987
41. Selkoe DJ, Bell DS, Podlisny MB, Price DL, Cork LC: Conservation of brain amyloid proteins in aged mammals and humans with Alzheimer's disease. Science 235:873-877, 1987
42. Gubler U, Hoffman BJ: A very simple and very efficient method for generating cDNA libraries. Gene 25:263-269, 1983
43. Neve R, Selkoe DJ, Kurnit DM, Kosik KS: A cDNA for a human microtubule-associated protein 2 epitope in the Alzheimer neurofibrillary tangle. J Molec Brain Res 1:193-196, 1986
44. Young RA, Davis RW: Efficient isolation of genes by using antibody probes. Proc Natl Acad Sci USA 80:1194-1198, 1983
45. Young RA, Davis RW: Yeast RNA polymerase II genes: isolation with antibody probes. Science 222:778-782, 1983
46. Huynh TV, Young RA, Davis RW: Constructing and screening cDNA libraries in λgt10 and λgt11, in Glover DM (ed): DNA Cloning Techniques: practical approach, vol 1. Oxford and Washington: IRL Press, 1985, pp. 49-78
47. Southern EM: Detection of specific sequences among DNA fragments separated by gel electrophoresis. J Molec Biol 98:503-517, 1975
48. Sims KB, Crandall JE, Kosik KS, Williams RS: Microtubule-associated protein 2 (MAP2) immunoreactivity in human fetal neocortex. Neurosci Abstr 12:212.5, 1986
49. Neve RL, Harris P, Kosik KS, Kurnit DM, Donlon TA: Identification of cDNA clones for the human microtubule-associated protein, tau, and chromosomal localization of the genes for tau and microtubule-associated protein 2. Molec Brain Res (in press)
50. Harper ME, Saunders GF: Localization of single copy DNA sequences on G-banded chromosomes by in situ hybridization. Chromosoma 83:431-439, 1981
51. Donlon TA, Litt M, Newcomb SR, Magenis RE: Localization of the restriction fragment length polymorphism D14S1 (pAW-101) to chromosome 14q32.1-32.2 by in situ hybridization. Am J Hum Genet 35:1097-1106, 1983
52. Feinberg AP, Vogelstein B: A technique for radiolabeling DNA restriction endonuclease fragments to high specific activity. Anal Biochem 18:5294-5299, 1979

53. Benton WD, Davis RW: Screening lambda-gt recombinant clones by hybridization to single plaques in situ. Science 196:180-182, 1977
54. Drubin DG, Caput D, Kirschner MW: Studies on the expression of the microtubule-associated protein, tau, during mouse brain development with newly isolated complementary DNA probes. J Cell Biol 98:1090-1097, 1984
55. Sharp PA: Conversion of RNA to DNA in mammals: Alu-like elements and pseudogenes. Nature 301:471-472, 1983

UBIQUITIN IN ALZHEIMER AND OTHER NEURODEGENERATIVE DISEASES

George Perry, Valeria Manetto, and Paul Mulvihill

Division of Neuropathology, Institute of Pathology, Case Western Reserve University, Cleveland, Ohio 44106

INTRODUCTION

The majority of efforts in understanding the abnormal neuronal inclusions found in some neurodegenerative diseases are directed toward analyzing their chemical nature. As discussed in other chapters, this analysis has depended primarily on indirect techniques e.g. immunocytochemistry and structural studies, which lack the quantitation and rigor that direct biochemical analysis can provide. Although these studies have demonstrated the presence of cytoskeletal elements in several inclusions, they have not provided an answer regarding their major constituent. Another approach in the study of these inclusions is the elucidation of the mechanism underlying the formation of abnormal polymers; that is, the processes responsible for transformation of components, probably normally present in neurons, into structures unique to these diseases.

Recently, it has been reported that ubiquitin, a protein important in regulating nonlysosomal proteolysis, is present in the neurofibrillary tangles (NFT) of Alzheimer disease (1,2). In this chapter we will consider the possible importance of this finding, specifically regarding one of the mechanisms that might be responsible for NFT formation and whether ubiquitin is associated with inclusions found in other neurodegenerative diseases.

Ubiquitin and Proteolysis

Ubiquitin, a 76 amino acid protein that is completely conserved from *Drosophila* to man, can be added to free amino groups of acceptor proteins by a unique posttranslational modification. Covalent conjugation of ubiquitin to the acceptor protein often marks the protein for subsequent proteolysis by a non-lysosomal pathway (3). A recent report has demonstrated that ubiquitin also contains an endogenous protease activity (4). Additionally, ubiquitin exists as a stable conjugate to several proteins including histones (5), lymphocyte homing receptor (6) and the receptor for platelet derived growth factor (7). The sequence conservation, presence of ubiquitin conjugated to several proteins having regulatory functions, and association with proteolysis indicates that ubiquitin probably plays a multifunctional role in directing cellular metabolism.

Presence of Ubiquitin in Alzheimer NFT

Using affinity purified antibodies to ubiquitin, Mori et al (1) and Perry et al (2) demonstrated ubiquitin in NFT, in neurites surrounding senile plaques and in altered neurites in the neuropil of Alzheimer disease. In tissue from aged neurologically normal individuals, only scant neurites and NFT, findings consistent with normal aging, were stained by antibodies to ubiquitin. Additionally, the antibodies recognized dot-like structures throughout the white matter in control and Alzheimer disease brains.

Analysis of peptide sequences obtained from paired helical filaments (PHF) (1) and immunocytochemistry of isolated PHF (1) have demonstrated the presence of ubiquitin. Further, when PHF are loaded onto SDS-polyacrylamide electrophoretic gels and subsequently immunoblotted, the primary band recognized by antibodies to ubiquitin is excluded from the stacking gel (2). Complementary results were obtained with the characterization of antibodies raised to detergent isolated PHF since several of these monoclonal antibodies and antisera recognize ubiquitin epitopes (1,8). Therefore, PHF, and possibly other structures characteristic of the Alzheimer brain, contain ubiquitin. Further, these results indicate that ubiquitin is an integral component of PHF that probably exists principally as a conjugate to an as yet unidentified protein or proteins.

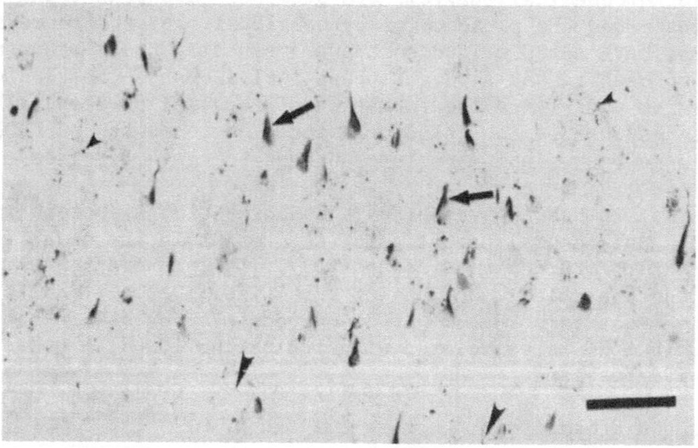

Figure 1. Ubiquitin immunoreactivity is associated with several pathological features of the Alzheimer brain, including neurofibrillary tangles (arrows), altered neurites (large arrowheads), and spherical structures (small arrowhead). Peroxidase-antiperoxidase. Light micrograph. X 115. Scale bar = 100 μm.

Ubiquitin is Present in Other Neuronal Inclusions of Degenerative Diseases

Antibodies to ubiquitin have been used to immunostain sections obtained at autopsy from patients having Pick and Parkinson diseases, progressive supranuclear palsy, central chromatolysis and amyotrophic lateral sclerosis. The fibrillary accumulations of central chromatolysis (9) and ALS (10) are composed of normal 10 nm neurofilaments; the others, although varying in morphology, primarily contain 12-18 nm straight filaments, which are not present in normal neurons. The affinity purified antibody to ubiquitin recognized the hallmark neuronal inclusions in each of the diseases containing abnormal filaments, but not the inclusions principally composed of neurofilaments. Therefore, the association of ubiquitin with intraneuronal inclusions is not specific to the NFT but apparently is associated with those inclusions characterized by morphologically abnormal filaments.

Figure 2. Antibodies to ubiquitin recognize "dot-like" structures throughout the white matter of brains obtained from Alzheimer and control cases. Peroxidase-antiperoxidase. Light micrograph. X 115. Scale bar = 100 μm.

Which Protein is Ubiquitinated?

The identity of the ubiquitinated protein(s) in these inclusion has not been established. The elegant study by Mori et al (1) defined ubiquitin sequences present in PHF; however, the proteolytic steps necessary to generate the peptides used in that analysis detach ubiquitin from its acceptor. Although the identity of the acceptor protein or proteins are unknown, two candidates are the microtubule associated protein tau and the high molecular weight neurofilament subunit (NFH). The structures recognized by antibodies to tau in sections from cases of Alzheimer and Pick diseases and PSP are similar to those stained by antibodies to ubiquitin, with the exception of the dot-like structures in the white matter and spherical structures which are not recognized by antibodies to tau.

Evidence to the contrary, however, is found in the Lewy body which contains ubiquitin epitopes but lacks any identified epitopes in common with tau (11). Since all of these inclusions contain NFH epitopes, there is a correspondence between presence of ubiquitin and NFH, but this correlation does not indicate that NFH is necessarily the acceptor. Identification of the acceptor protein(s), which must await further chemical analysis of the ubiquitin conjugates, is vital to understanding the role ubiquitin plays in the formation of these inclusions.

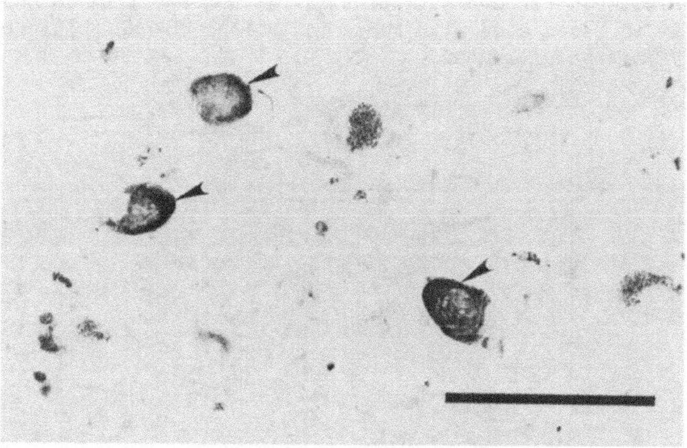

Figure 3. NFT found in the brain stem of a progressive supranuclear palsy case although composed of straight filaments, rather than PHF, are recognized by antibodies to ubiquitin (arrowheads). Immunostained with an affinity purified antiserum to ubiquitin. Light micrograph. Peroxidase-antiperoxidase. X 290. Scale bar = 100 µm.

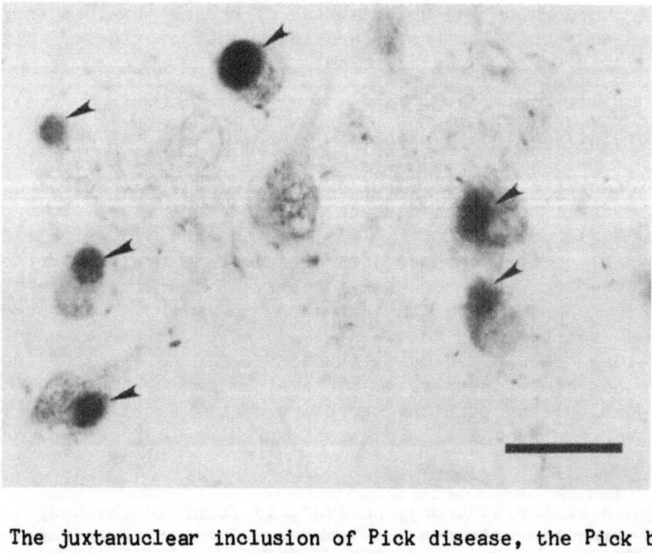

Figure 4. The juxtanuclear inclusion of Pick disease, the Pick body, is recognized by antibodies to ubiquitin. In this light micrograph, Pick bodies (arrowheads) are intensely immunostained by an affinity purified antiserum to ubiquitin. Peroxidase-antiperoxidase. Magnification X 825. Scale bar = 20 µm.

Significance of Ubiquitin in Neuronal Inclusions: Conclusions and a Hypothesis

Among the various interpretations one can make regarding the high density of ubiquitin in the inclusions are the following two: 1) the inclusions are ubiquitinated as the cellular response to the inclusion and/or 2) ubiquitin plays a role in processing of constituents leading to PHF formation. Both these interpretations suggest a deficit in the ubiquitin-mediated proteolytic pathway, since the ubiquitin conjugates present are apparently stable. Although we cannot rule out ubiquitin conjugation as a generalized response to the abnormal proteins found in these inclusions, it is possible that the defect in the ubiquitin system is instrumental in the formation of these filamentous inclusions.

The primary role of proteolysis in amyloid fibril formation may be analogous to and serve as the model for the processes that lead to the formation of PHF and other abnormal filaments found within neuronal inclusions. Recent studies by Kang et al (12) have demonstrated that the amyloid fibrils of senile plaques and cerebral vascular angiopathy are a product of the incomplete proteolysis of a larger precursor protein, yielding a small peptide termed A4. The 8-10 nm straight filaments of cerebral amyloid, although structurally distinct from PHF, share the same beta pleated structure, as demonstrated by X-ray diffraction studies (13). Whether the filaments within other neuronal inclusions are also in a beta pleated configuration and whether the beta pleated configuration is indicative of a common mode of formation remains to be established.

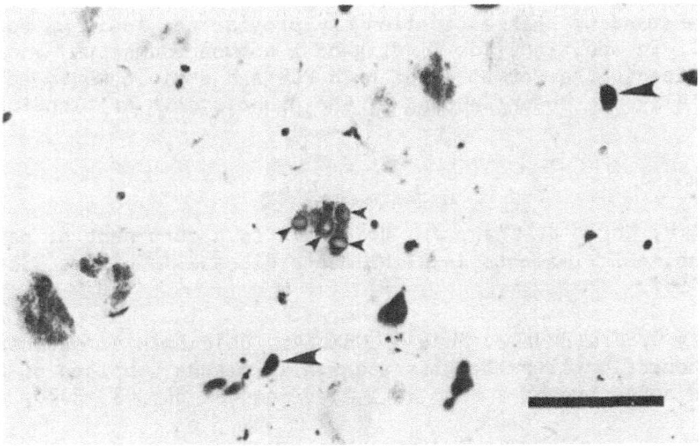

Figure 5. Lewy bodies, the hallmark neuronal inclusion of pigmented neurons in the substantia nigra and locus coeruleus of Parkinson disease cases, are intensely immunostained by antibodies to ubiquitin (small arrowheads). Additionally, antibodies to ubiquitin recognized some altered neurites (large arrowheads). Light micrograph. Peroxidase-antiperoxidase. Magnification X 360. Scale bar = 50 μm.

Although the chemical relationship between intraneuronal filamentous inclusions and amyloid is not known, one could hypothesize that both might result from a similar process of incomplete proteolysis. In Alzheimer disease, the presence of PHF within altered neurites surrounding senile plaques and within other altered neurites, as well as in NFT, bears on this hypothesis. Altered neurites of Alzheimer disease are intensely stained with antibodies to ubiquitin. In axons, which, along with dendrites, constitute neurites, the cytoskeletal components are assembled into polymers prior to being transported down the axon to the terminal (14). Physiologically, proteolysis at the axon terminal is responsible for recycling axonal components by breaking them down to free amino acids (15). Therefore, disruption of proteolytic processing in neurites may contribute to PHF formation. Support for a neuritic role in PHF formation is found in identified PHF components, phosphorylated NFH and tau, both of which are normally restricted to neurites. Partial, and presumably abnormal ubiquitin dependent proteolysis of NFH and tau, as well as possibly unidentified neuritic components (16), would provide the building blocks for new polymers. The presence of ubiquitin within the altered neurites of Alzheimer disease further suggests that ubiquitin may play a role in the degradation/regulation of the cytoskeleton in normal neurons. The role neurites might play in the formation of Pick bodies and Lewy bodies and NFT of progressive supranuclear palsy is more ambiguous. Although antibodies to ubiquitin recognize altered neurites in the brain stem of Parkinson and progressive supranuclear palsy cases, their relationship to neuronal inclusions has not been established. Possibly the role of proteolysis in the formation of each of these inclusions is not identical, since the etiology of each of these diseases is probably distinct.

Although the role ubiquitin plays in Alzheimer disease or other neurodegenerative diseases is not established, it has opened the way to new insights concerning the cellular basis and mechanism responsible for the formation of novel filaments. Alteration of protein metabolism in neurites is probably not the primary lesion of any of these diseases. Yet, understanding this alteration may provide new insights concerning etiology. In addition, the finding of a common mechanism, incomplete proteolysis, in the formation of both PHF and amyloid would advance substantially the understanding of the pathogenesis of Alzheimer disease.

REFERENCES

1. Mori H, Kondo J, Ihara Y. Ubiquitin is a component of paired helical filaments in Alzheimer's disease. Science 235:1641-1644, 1987.

2. Perry G, Friedman R, Shaw G, Chau V. Ubiquitin is detected in neurofibrillary tangles and senile plaque neurites of Alzheimer disease brains. Proc Natl Acad Sci USA 84:3033-3036, 1987.

3. Finley D, Varshavsky A. The ubiquitin system: functions and mechanisms. Trends Biochem Sci 11:343-347, 1985.

4. Fried VA, Smith HT, Hildebrandt E, Weiner K. Ubiquitin has intrinsic proteolytic activity: Implications for cellular regulation. Proc Natl Acad Sci USA 84:3685-3689, 1987.

5. Busch H, Goldlenopf IL. Ubiquitin-protein conjugates. Mol Cell Biochem 40:173-187, 1981.

6. Siegelman N, Bond MW, Gallatin WM, St John T, Smith HT, Fried VA, Weissman IL. Cell surface molecule associated with lymphocyte

homing is a ubiquitinated branched-chain glycoprotein. Science 231:823-829, 1986.

7. Yarden Y, Escobedo JA, Kuang WJ, Yang-Feng TL, Daniel TO, Tremble PM, Chen EY, Ando ME, Harkins RN, Francke U, Fried VA, Ullrich A, Williams LT. Structure of the receptor for platelet-derived growth factor helps define a family of closely related growth factor receptors. Nature 323:226-232, 1986.

8. Perry G, Mulvihill P, Manetto V, Fried V, Smith H, Grundke-Iqbal I, Iqbal K, Autilio-Gambetti L, Gambetti P. Ubiquitin in Alzheimer paired helical filaments: a study with monoclonal antibodies. Soc Neurosci. Abstract. (submitted).

9. Moss TH, Lewkowicz SJ. The axon reaction in motor and sensory neurons of mice studied by a monoclonal antibody marker of neurofilament protein. J Neurol Sci 60:267-280, 1983.

10. Hirano A. Aspects of the ultrastructure of amyotrophic lateral sclerosis. In: LP Rowland (Ed), Human Motorneuron Disease, Raven Press, New York, 1982, pp 75-88.

11. Galloway P, Grundke-Iqbal I, Autilio-Gambetti L, Gambetti P, Perry G. Neuronal cytoskeletal involvement in Lewy body formation. J Neuropathol Exp Neurol 46:374, 1987.

12. Kang J, Lemaire HG, Unterbeck A, Salbaum JM, Masters CL, Grzeschik KH, Malthaup G, Beyreuther K, Muller-Hill B. The precursor of Alzheimer's disease amyloid A^4 protein resembles a cell-surface receptor. Nature 325:733-736, 1987.

13. Kirschner D, Abraham C, Selkoe D. X-ray diffraction from intraneuronal paired helical filaments and extraneuronal fibers in Alzheimer disease indicates cross-β conformation. Proc Natl Acad Sci USA 83:503-507, 1986.

14. Lasek RJ, Garner JA, Brady ST. Axonal transport of the cytoplasmic matrix. J Cell Biol 99:212-221, 1984.

15. Schlaepfer WW, Lee C, Lee WMY, Zimmerman UPJ. An immunoblot study of neurofilament degradation in situ and during calcium-activated proteolysis. J Neurochem 44:502-509, 1985.

16. Perry G, Friedman R, Kang DH, Manetto V, Autilio-Gambetti L, Gambetti P. Antibodies to the neuronal cytoskeleton are elicited by Alzheimer paired helical filaments fractions. Brain Res (in press).

NEURONAL FIBRILLAR CYTOSKELETON AND ENDOMEMBRANE SYSTEM ORGANIZATION IN ALZHEIMER'S DISEASE

Mark Ellisman[1]*, Rama Ranganathan*, Thomas Deerinck*, Stephen Young*, Robert Terry and Suzanne Mirra[2]

*Laboratory for Neurocytology M008, Department of Neurosciences, UCSD, La Jolla, CA 92093 and[2] Veterans Administration Medical Center and Emory University School of Medicine, Atlanta, GA 30033

ABSTRACT

We examined the structural organization of cytoskeletal components and membrane systems in neurons of well preserved biopsy material from Alzheimer's disease (AD). Information was obtained from thin sectioned material using conventional electron microscopy and from thick sections with the high voltage electron microscope. Stereo viewing and computer assisted serial reconstruction techniques were employed to visualize three-dimensional relationships among cytoplasmic components of cortical neurons.

We observed several interesting ultrastructural features in AD neurons. These include associations of paired helical filaments (PHF) with the membranes of the nuclear envelope and with ribosomes, differences between the distribution of the Golgi apparatus in neurons containing paired helical filament bundles as compared to neurons without these filaments, and abnormalities in the microtubules of neuronal processes in the vicinity of neuritic plaques.

INTRODUCTION

Disruptions in the fibrillar cytoskeleton of AD neurons are well documented (Wisniewski and Terry, 1976). More recently, alterations in the location and nature of at least one microtubule-associated protein, tau, have been noted (Wood et al., 1986; Kosik et al., 1986; Grundke-Iqbal et al., 1986). In AD, tau, usually found in axons, is widely broadcast in association with PHF. PHF are found in perikarya, proximal dendrites, occasionally within myelinated axons and very commonly within components of neuritic plaques. The tau appears to be associated with the PHF in a highly phosphorylated state (Wood et al., 1986; Grundke-

[1]To whom correspondence should be addressed; supported by NIH NS14718, AG05131, RR00592 and the NSF; [2]Supported by VA Merit Award #5747-004.

Iqbal et al., 1986). Apparently, some epitopes of neurofilament peptides in PHF also appear to be extensively phosphorylated (Sternberger et al., 1985; Cork et al., 1986). Thus, both structural and biochemical studies indicate alterations in major components of the microtubule and neurofilament-based cytoskeleton. It is now well established that both of these filament systems participate in the regulation of cell form and the maintenance of the spatial organization of the major cytoplasmic membrane systems. Not yet understood, however, is the extent to which the positional and functional relationships normally shared by elements of the endomembrane system are disrupted by this pathological condition.

New information about the organization of the neuronal endomembrane system (Lindsey and Ellisman, 1985a,b,c) the association of the Golgi complex with the initiation of axonal transport (Hammerschlag et al., 1982) and the membrane systems of the axon (Ellisman and Lindsey, 1982) is available. However, correlation of this information with data from previous studies of the pathology of AD neurons is difficult. Although historically of great interest, previous work on the effects of aging on neuronal Golgi apparatus structure are varied. Early light microscopic studies claim age-related changes in the Golgi apparatus (Golgi, 1898, 1899, 1900; Andrew, 1939; Gatenby and Moussa, 1951; Strehler, 1976). No significant changes were reported in the only EM study in which the morphology of the Golgi in aged neurons was mentioned (Hasan and Glees, 1973). This report, however, addresses the appearance of Golgi stack lamellae in thin section rather than the extent, size or distribution of the Golgi complex within the cell. To our knowledge there are no studies dealing with the morphology and distribution of the Golgi apparatus or other components of the endomembrane system in AD.

The objectives of this work were to further characterize the distribution of the fibrillar components of the cytoskeleton and to evaluate the status of major elements in the neuronal endomembrane system within AD neurons. Well-preserved samples of biopsy material of cerebral cortex from patients with neuropathologically confirmed AD were studied. We used conventional transmission EM (CTEM) and high voltage EM (HVEM) as well as computer assisted serial section reconstruction techniques to gather new information about structural relationships in neurofibrillary tangle-bearing neurons. These findings were compared to those in apparent tangle-free neurons from the same biopsy materials.

METHODS

The Collection of Serial Sections, the Staining and the Conventional Electron Microscopy

Epoxy embedded blocks of tissue were sectioned using a specially equipped diamond knife in which the boat had been modified to resemble a very small marina. Fine dissecting needles lined the back of the boat and provided slips into which ribbons of serial sections were sequentially docked. After a complete series of 100 1/4μm sections were safely harbored, the sections were individually led onto numbered formvar filmed slot grids using a platinum transfer-loop technique. The sections were thus individually collected and kept in order throughout subsequent processing. Serial sections were uniformly stained by sequential immersion in 2% aqueous uranyl acetate and SATO lead stain for 1 minute each. Upon drying, sections were coated on each side with approximately 200A of carbon for dimensional stability.

Microscopy was conducted using a JEOL 100CX equipped with a eucentric goniometric stage and a specimen holder capable of rotation to adjust the orientation of the specimen field on the micrograph. The position of a specific cell is first established for a section near the middle of a series. Criteria for selection of cells varied depending upon objectives of the comparisons to be made but, in general, when in the middle of the series we chose cells in which the cell nucleus presented a large profile and a portion of the nucleolus had been included. This ensured inclusion of the cell body and segments of its major arborizations in the serial sectioning for reconstruction.

After the micrographs were developed and numbered, they were carefully examined for quality. If additional information was needed in a given plane of the series, the section was reshot so that additional micrographs represented a montage of the area surrounding the original field of interest or a record of detail at higher magnification. This, for example, allowed us to track the major arborizations (dendrites and axons) out of the field containing the cell soma. Each series of micrographs was then printed on 11"x14" sheets of resin-coated photographic paper with a point-source enlarger.

Digitization of the Micrographs

We used a computer system for serial section reconstruction developed by one of us (Young et al., 1987). Although the basic software is now available upon request as a service of the Boulder Laboratory for HVEM and the NIH Division of Research Resources, we used a special version for this work. This was created in order to accommodate the large data files required when keeping track of the major cytoplasmic organelles within an entire cell through 100 sections. The computer software is very "user-friendly" and utilizes a computer in the IBM-PC series with a 16 color, 640x400 pixel graphics display card and monitor. User interaction with the system is accomplished through a keyboard and a high resolution graphics tablet. Textual information is displayed on a second monitor (monochrome).

Before entering the detailed location of cell boundaries and organelles from each section, a set of fiduciary marks were established on each 11"x14" print. This was done by laying one micrograph on top of its neighbor in the series over a very bright light box and moving one over the other until key structural details lined up on both. A pin was then pushed through both in at least six places to mark correspondence, and these markings were carried through the entire series. These fiduciary marks were numbered 1-6 identically on each plane and served as alignment coordinates for the reconstruction. Structures to be entered in the reconstruction were first outlined with a colored, sharp marking pen (ethanol soluble). The fiduciary marks and color coded ultrastructural details were then entered section by section by tracing on the digitizing tablet.

Preparation of Sections for HVEM

Thick sections (1/2-1μm) were cut from the same blocks of tissue from which serial sections were cut. These were stained with lead and uranium salts as described previously (Ellisman and Porter, 1980; Lindsey and Ellisman 1985a,b,c). Sections were examined at 750 KeV and 1 MeV using the HVEM located in Boulder, CO. Stereopair micrographs were taken of interesting areas with useful tilt differences determined according to the magnification, specimen thickness and relative dimensions of the structures of interest. The degree of tilt, magnification, and accelerating voltage for each micrograph are noted in the figure captions.

OBSERVATIONS

In these investigations we have examined the distribution of the major classes of organelles and pathologic changes in neurons of biopsy tissue from the brains of AD patients. Although abnormalities in the fibrillar cytoskeleton of AD neurons have been well documented, little is known about the three-dimensional distribution of the fibrillar cytoskeleton in affected neurons and its relation to the major organelle systems. To obtain more information about these relationships, we examined the locations and distributions of cell boundaries, nuclear boundaries, nucleoli, rough endoplasmic reticulum, Golgi apparatus, paired helical filaments, lipofuscin (LF), basal bodies and cilia with three-dimensional reconstructions and thick-section stereo micrographs (Ellisman et al., 1986). General comparisons were made between cells of similar overall form and size but differing with respect to the type of inclusions present. For example when comparing cells with respect to the distribution of the Golgi apparatus, categories were defined as those containing PHF and LF, those containing large amounts of LF but no PHF, and those considered "normal", containing little LF and also without PHF.

The computer graphic reconstructions are illustrated by the two examples of Figures 1 and 2. These images are photographs of the computer color display of the reconstructed series rotated to various angles on the vertical axis. These stereo images were obtained by photographing reconstructions at tilt angles differing by 10° of rotation. One of these cells (Fig. 1) is a reconstructed neuron that contained a moderate amount of PHF (in red) which are noted in close proximity to the membranes of the nuclear envelope (in yellow). PHF were found in proximal dendritic branches. The small red-brown globe-like structure inside the nucleus is the nucleolus. In these reconstruction studies of PHF bearing neurons, the Golgi apparatus (in white) was found to be peripheralized relative to more normal appearing neurons (without PHF and with or without LF). The Golgi also appeared more globular and was not as completely extended into sheets as found in more normal neurons (see Fig. 2). Interestingly, the lipofuscin in this neuron (brown and green) is almost completely encapsulated by the PHF. Note that with the reconstruction graphic system employed, the reconstructions can be derived with or without any of the major subcategories of organelle depicted (and colors may also be modified). In the case of the lipofuscin, which in these images is partially obscured by the PHF, one simply has to view a version of the reconstruction without the PHF to expose the lipofuscin distribution.

Figure 2 is a reconstruction of a more normal-appearing neuron from nearby in the same biopsy serial section series as the neuron of Fig. 1. Although this neuron contained no recognizable PHF it did contain considerable lipofuscin (brown and green). The form of this neuron appears less distorted than that of the one reconstructed in Figure 1. It is easy to recognize that this is a cortical pyramidal neuron by the apical dendrite which extends out toward the viewer. Note the single cilium (characteristic of differentiated neurons *in situ* [Peters et al., 1976]) extending downward from the cell body in these images. The axon of this neuron is projecting out from the cell soma at about 11 o'clock. The Golgi apparatus of this neuron can be seen to encapsulate the nucleus and extend partly into the major dendritic arborizations.

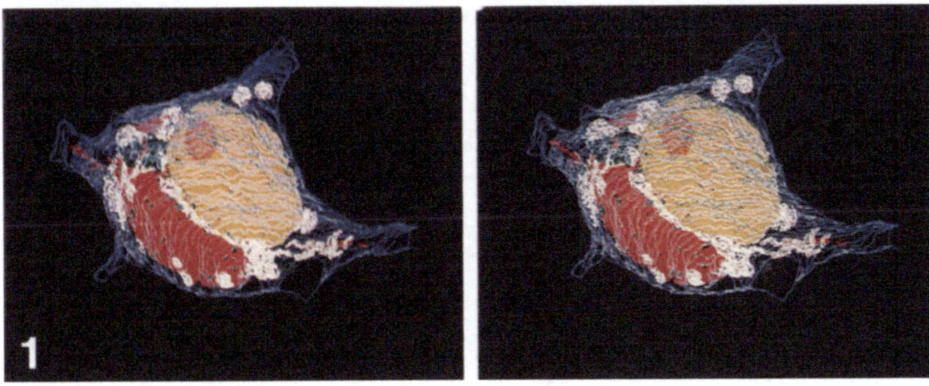

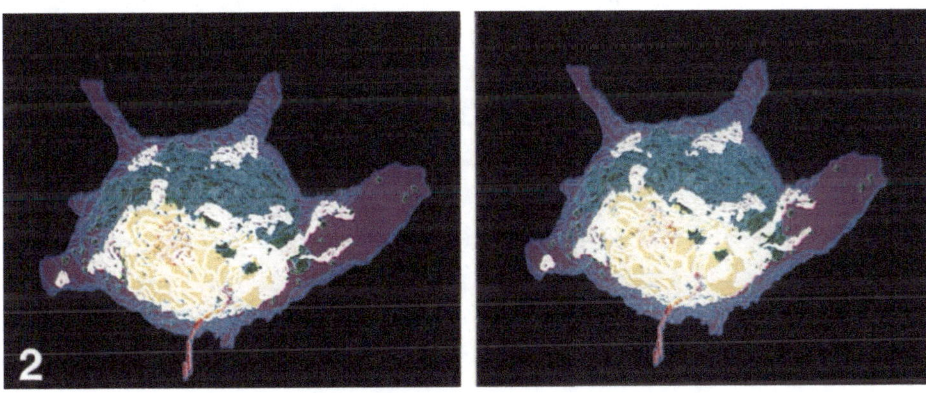

Figures 1 and 2 are examples of reconstructions from serial sections. These were obtained by employing the reconstruction methodology described in the text. For proper orientation these and the stereo pairs of Figures 5 and 6 are to be viewed with a two-lens stereo viewer. These computer graphic reconstructions are through the neuronal somata and initial portions of arborizations of two neurons from the same biopsy. Figure 1 is of a neuron that contained both PHF and LF. Figure 2 is a reconstruction of a neuron that contained only LF. A detailed description may be found in the text.

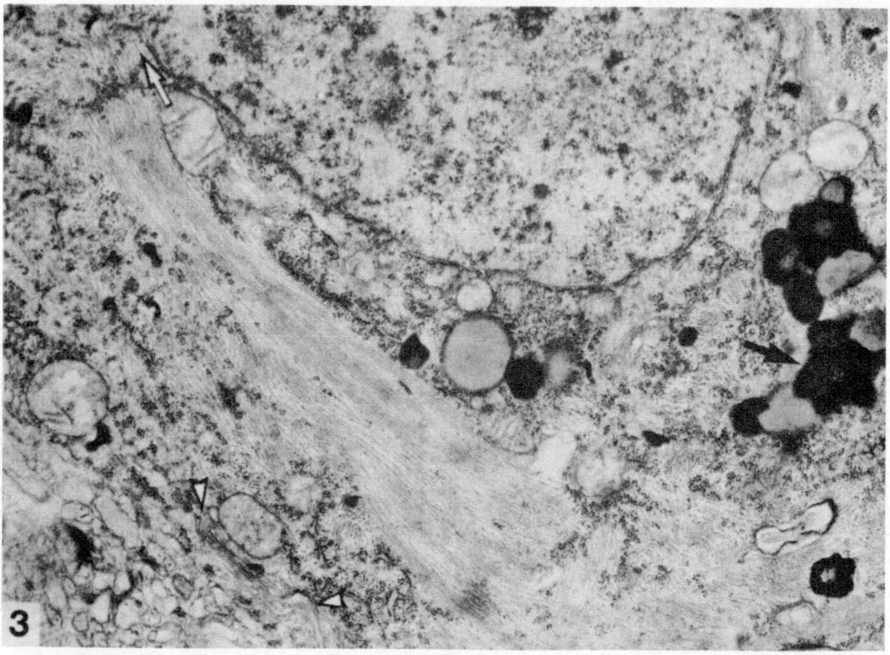

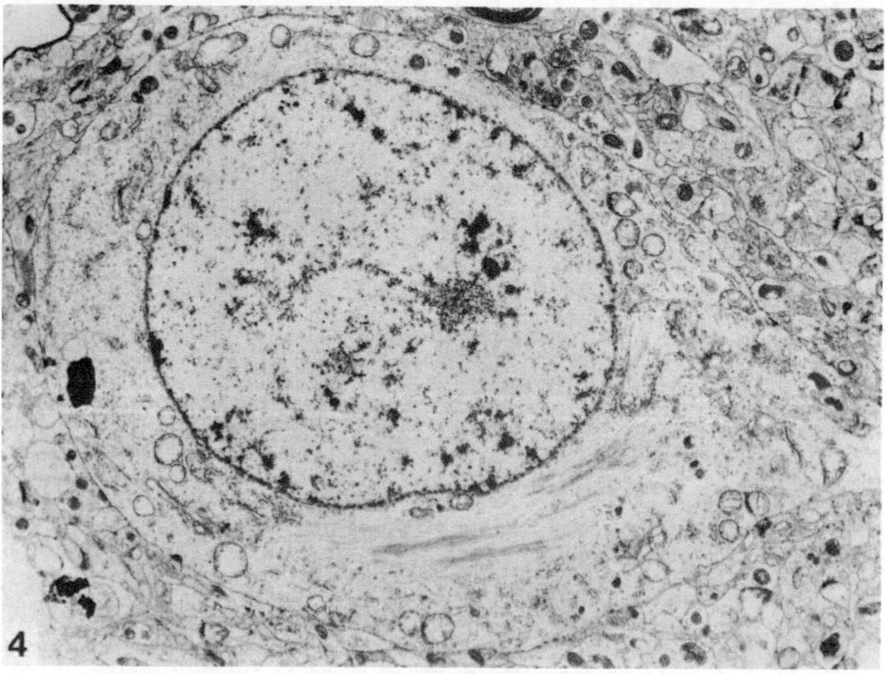

Figures 3 and 4 are relatively low power electron micrographs from sections of two neurons which exhibit different amounts of PHF inclusions. They are from different biopsies. Open arrow, PHF near nucleus; open arrowhead, Golgi; black arrow, LF; asterisk, RER. Fig. 3 magnification, 15,000X. Fig. 4 magnification, 6,000X.

As is known from previous ultrastructural studies of Alzheimer neurons, not all affected neurons contain equivalent amounts of PHF. It is generally assumed that neurons exhibiting smaller numbers of PHF represent earlier stages of a degenerative process. We have surveyed the cytoplasmic ultrastructure of neurons ranging from those with the cytoplasm nearly filled with PHF (Fig. 3) to those containing only sparse PHF (Fig. 4). Figures 3 and 4 are from different biopsies although variations in the amount of the PHF may be found among neurons of the same biopsy.

Figure 3 shows a relatively broad bundle of PHF closely approximating the membranes of the nuclear envelope (open arrow). A segment of the Golgi apparatus can be seen in an abnormal location near the plasma membrane (open arrowheads). In reconstructions of such neurons we found that domains of cytoplasm rich in LF also contained RER. In contrast, RER and LF appeared excluded from the PHF bundles. Note the relationship between the lipofuscin (closed arrow) and the PHF and rough endoplasmic reticulum (asterisk).

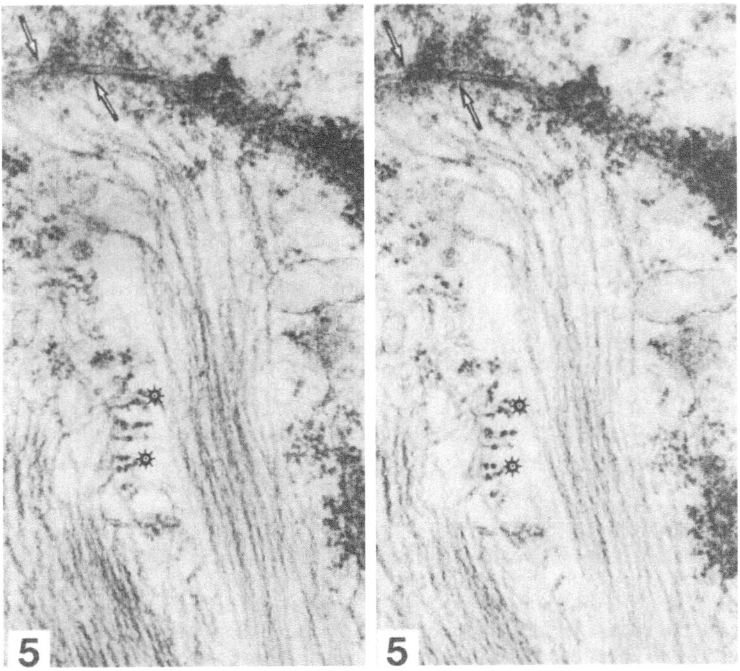

Figure 5 is a stereo pair which illustrates the "end-on" associations made between PHF and the membranes of the nuclear envelope (open arrows) and polyribosomes that are in association with the ends of some PHFs (asterisks). Magnification, 52,000X; section thickness, 1/4μ; 100 KeV; and 10° of tilt difference between images.

Very few microtubules were found in the cytoplasm of neurons exhibiting extensive PHF as in the reconstructed cell of Figure 1 or Figure 3. In contrast, cells like the one depicted in Figure 4 (taken from a biopsy in which many blocks contain neurons with fewer PHF) contain variable numbers of microtubules.

In general, PHF-laden neurons contain few to no cytoplasmic microtubules. In cells with scant to moderate amounts of PHF more microtubules were observed. Mixed complexes of PHF bundles and microtubules were occasionally seen in the cell bodies, but microtubule rich zones appeared to be separate from PHF-rich zones in most instances.

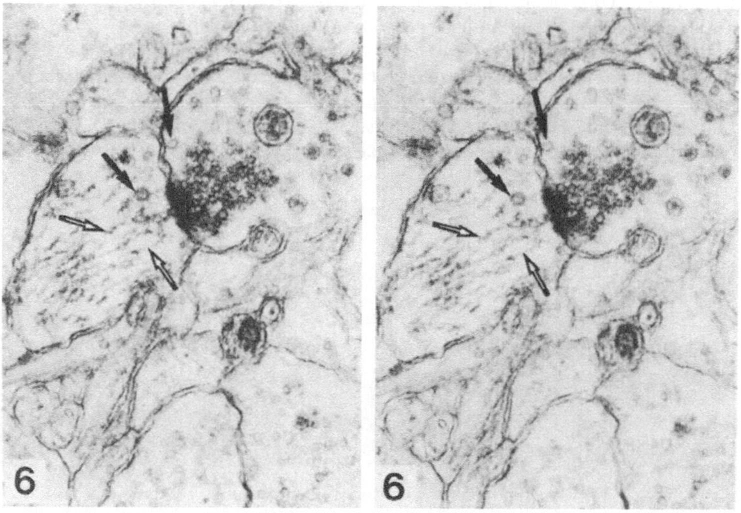

Figure 6 is from a slightly thicker section and illustrates an association of PHF (open arrows) with material just beneath the postsynaptic density. See text for details. Magnification, 31,000X; section thickness, 1/2μ; 750 KeV; and 10° of tilt difference between images.

In addition to the low magnification micrographs described above and used for cell reconstructions, high magnification stereo images revealed interesting associations between PHF and cell structures. Figures 5 and 6 are examples of such stereo pairs from both conventional and high voltage electron microscopy. Figure 5 illustrates the "end-on" associations observed between PHF and the membranes of the nuclear envelope (open arrows). The site of this association was first defined by following the filaments in the computer reconstruction and then choosing a subset of sections to examine in greater detail using the tilting stage of the conventional electron microscope operated at 100 KeV. The images are therefore from a cell that was reconstructed but were taken at higher magnifications. This set of micrographs also illustrates that polyribosomes may be seen in association with the ends of some PHF (asterisks).

We also examined the ultrastructure of axons and dendrites at synapses in the vicinity of neuritic plaques. We were interested in the morphology and associations of PHF and microtubules at these terminal arborizations. Figure 6 is a stereo pair of high voltage electron

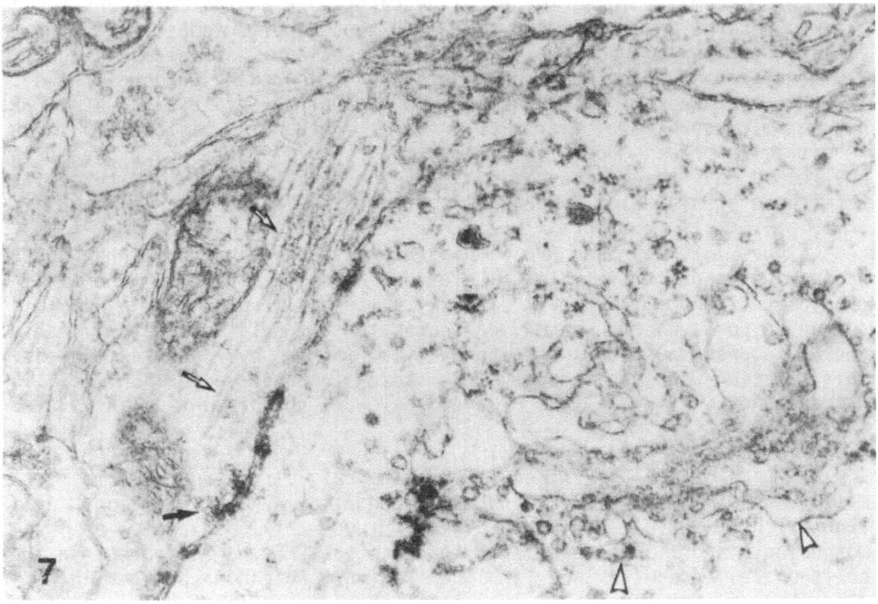

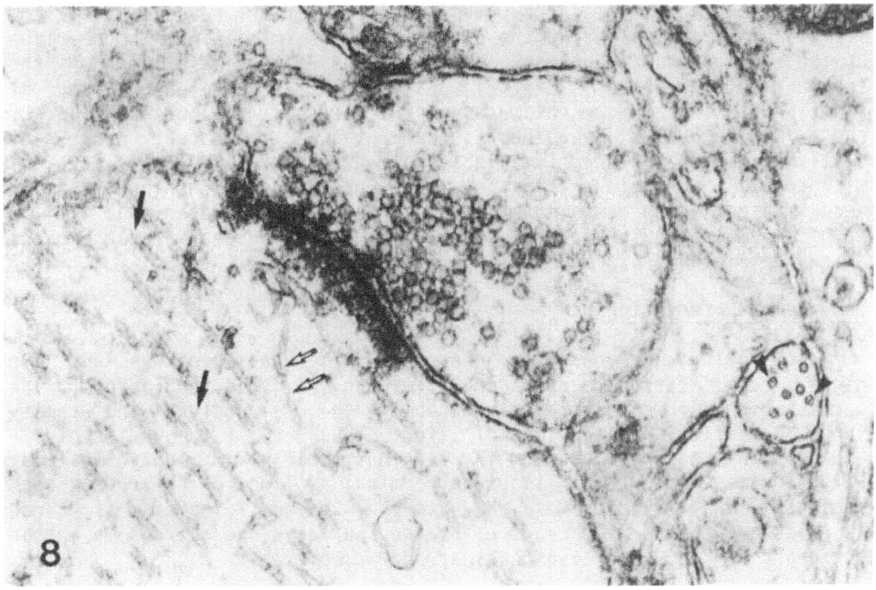

Figures 7 and 8 are high voltage electron micrographs which have been examined by us in 3-D but are not presented in stereo here. Figure 7 illustrates that PHFs are also found in association with presynaptic boutons. Figure 8 is from a nearby area in which few PHFs were apparent in the cells or cell processes. See text for details. Figure 7 magnification, 35,000X; Figure 8 magnification, 60,000X; section thicknesses were 1/2μ imaged at 750 KeV.

micrographs (750 KeV accelerating voltage) from a slightly thicker section (1/2μm). This image illustrates an association of PHF (open arrows) with material just beneath the postsynaptic density of what appears to be a functional synaptic connection. Note also the coated vesicles present in both pre and postsynaptic processes (closed arrows). In this instance the postsynaptic component of the synapse contains PHF but no microtubules. Microtubules are frequently visible in such dendritic arbors. Speculation on the viability of such synapses is buttressed by the presence of coated vesicles that indicate the occurrence of biological processes at the time of fixation.

Figure 7 illustrates PHF within a presynaptic bouton (PHF, open arrows; synaptic vesicles, closed arrow). This PHF-containing presynaptic bouton is applied directly to the cell body of a normal-appearing postsynaptic neuron. A segment of the Golgi apparatus in this neuron is present in this section and is marked by open arrowheads. This section was approximately 1/2μm thick and observed at 750 KeV. It was examined in 3-D but like Figure 8 below is not presented in stereo pairs here.

As indicated above we noted atypical ultrastructure in the microtubules of otherwise normal appearing synaptic regions near abnormal neurites of neuritic plaques. Figure 8 is an example of such morphology and is from an area in which few PHFs were apparent in the cells or cell processes near a plaque region. Note however that in the dendrite of the synaptic apposition there is at least one suspicious PHF-like structure (open arrows). Close inspection reveals that the microtubules indicated by the closed arrows terminate within the plane of the section. Although many are continuous from one surface of the section to the other, careful analysis in 3-D revealed that several start or stop within the thickness of the section. This observation of short microtubules suggests recurrent and abnormal cycles of depolymerization and repolymerization. Alternatively, this atypical microtubule morphology may be the result of immersion fixation necessitated when processing biopsy material. Note however that similar atypical microtubular profiles are common within peri-plaque regions which contain numerous abnormal neurites.

Summary of Observations

The following is based on the complete reconstructions and the higher magnification stereoscopic imaging of selected regions. The accumulations of PHF partially encapsulate the cell nucleus and ramify into dendritic arborizations of affected neurons. Some bundles of PHF end within the cell body in regions rich in polyribosomes. Polyribosomes are seen along the terminal portions of the PHF in such regions. Many bundles of PHF can be traced to sites of close association with the membranes of the nuclear envelope where some appear to insert in the vicinity of nuclear pore complexes.

The density of somal microtubules visible in the PHF-containing cells ranged from few to none. Numerous microtubules were visible, however, in the LF rich cells without PHF and in the "normal" cells. The RER was closely associated with the LF but appeared excluded from the PHF bundles.

In the PHF-containing cells, portions of the Golgi apparatus were more peripheral than in normal appearing neurons., In contrast, the Golgi complex was more normally distributed in both the "LF only" and "normal" cells. Atypical morphology was noted in some of the microtubules from these specimens.

DISCUSSION

These observations indicate that several changes in the cytoplasmic organization of PHF-containing neurons appear to have occurred. These have specific implications for the functional status of these neurons and also suggest several loci as possible sites of primary defects resulting in this abnormality. For instance: Golgi apparatus function may be impaired and thus contribute to cell death; affected microtubules in neurites may effect the delivery and retrieval of materials to and from the distal portions of neuronal arborizations; and, as suggested by Suzuki and Terry (1967), axonal transport may be impaired.

The current literature documenting observations on AD tissue and many of the chapters in this volume indicate that there is still considerable controversy as to the nature of the filamentous material which accumulates within these abnormal cells. It is our contention that a 3-dimensional approach to the ultrastructural pathology in AD might add to our understanding of the nature of these changes. From our preliminary study it is clear that, although the neurofibrillary tangle occupies a considerable volume within the cytoplasm, the remainder of the cell still contains the usual cellular organelles although microtubules appear to be reduced in number. Despite the persistence of most functional complexes, e.g. Golgi apparatus, they appear to be redistributed and there is a subtle change in the shape of the Golgi apparatus as well as a decrease in the number of somal microtubules found in PHF-containing neurons.

Although more quantitative and detailed studies need to be conducted, one has the impression from these biopsies that the larger the amount of PHF inclusion, the more peripheralized the Golgi apparatus becomes and the more scarce microtubules become. We are inclined to speculate that the microenvironment of neurons and especially of the neuron's terminal arbor may be unhealthy for microtubules. This may promote a type of retrograde degeneration of the parent cell body that includes disruption of the somal cytoskeleton.

Findings of a decentralization of the Golgi complex in neurons containing few if any microtubules is consistent with observations on the relationship between these structures in other cell types. For example, Singer and co-workers (Rogalski et al., 1984a,b) have recently shown that the Golgi apparatus depends upon microtubules to maintain its position in the cytoplasm of cultured cells. The redistribution of the Golgi apparatus in PHF-containing neurons therefore may be a secondary consequence of other factors which destabilize microtubules.

By conducting detailed electron microscopic examination of biopsies containing neurons in very early stages of the degenerative process we expect to gain further insight into such possibilities. We have already observed evidence for certain early appearing abnormalities such as altered structure and distribution of microtubules, altered Golgi distribution, and simplification of the dendritic arbor. Certain of these alterations in cellular functioning may contribute toward the sequellae of neuronal degeneration and cortical isolation noted in other studies (Terry et al., 1981; Morrison et al., 1983; Campbell et al., 1985; Mann et al., 1985; Pearson et al., 1985; Rogers and Morrison, 1985; Lewis et al., 1986a,b). More information about the early fates of subcellular structures which participate in major cellular processes such as axonal transport (microtubules), protein synthesis (the endoplasmic reticulum) and glycosylation (the Golgi apparatus) needs to be obtained. We would expect these investigations to elucidate the nature of changes in such structures.

REFERENCES

Andrew, W. (1939) The Golgi apparatus in the nerve cells of the mouse from youth to senility. Amer. J. Anat. 64: 351-375.

Campbell, M.J., Lewis, D.A. and Morrison, J.H. (1985) The immunohistochemical distribution of somatostatin 28 (SS28) and somatostatin 28 (1-12) (SS28(1-12)) in monkey prefrontal (PFC) and temporal (TC) cortical regions. Soc. Neurosci. Abstr. 11: 679.

Cork, L.C., Sternberger, N.H., Sternberger, L.A., Casanova, M.F. and Struble, R.G. and Price, D.L. (1986) Phosphorylated neurofilament antigens in neurofibrillary tangles in Alzheimer's disease. J. Neuropath. Exper. Neurol. 45: 56-64.

Ellisman, M.H. and Lindsey, J.D. (1982) Organization of axoplasm: membranous and fibrillar components possibly involved in fast neuroplasmic transport. In Axoplasmic Transport (ed. D. Weiss), Raven Press, NY, pp. 55-63.

Ellisman, M.H., and Porter, K.R. (1980) Microtrabecular structure of the axoplasmic matrix: visualization of cross-linking structures and their distribution, J. Cell Biol., 87:464.

Ellisman, M.H., Ranganathan, R., Deerinck, T.J., Young, S.J., Terry, R. and Mirra, S. (1986) Fibrillar cytoskeleton and endomembrane system organization in Alzheimer neurons. Soc. Neurosci. Abstr. 12: 268.

Gatenby, J.B. and Moussa, T.A.A. (1951) The neurone of the human autonomic system and the so-called 'senility pigment'. J. Physiol. 114: 252-254.

Golgi, C. (1898) Intorno alla struttura delle cellule nervose. Boll. Soc. Med. Pavia 1: 3-16.

Golgi, C. (1899) Di nuovo sulla struttura delle cellule nervose dei gangli spinali. Arch. Ital. Biol. 31: 273-180.

Golgi, C. (1900) Intorno alla struttura delle cellule nervose della corteccia cerebrale. Verhand. d. Anat. Gesell. 14: 164-176.

Grundke-Iqbal, I., Iqbal, K., Tung, Y.C., Quinlan, M., Wisniewski, H.M. and Binder, L.I. (1986) Abnormal phosphorylation of the microtubule associated protein tau in Alzheimers cytoskeletal pathology. Proc. Natl. Acad. Sci. U.S.A. 83: 4913-4917.

Hammerschlag, R., Stone, G.C., Lindsey, J.D. and Ellisman, M.H. (1982) Evidence that all newly synthesized proteins destined for fast axonal transport pass through the Golgi apparatus. J. Cell Biol. 93: 568-575.

Hasan, M. and Glees, P. (1973) Ultrastructural age changes in hippocampal neurons, synapses and neuroglia. Exp. Geront. 8: 75-83.

Kosik, K.S., Joachim, C.L. and Selkoe, D.J. (1986) Microtubule-associated protein, tau, is a major antigenic component of paired helical filaments in Alzheimer's disease. Proc. Natl. Acad. Sci. U.S.A. 83: 4044-4048.

Lewis, D.A., Campbell, M.J., Huntley, G.W., Benson, D.L., Terry, R.D. and Morrison, J.H. (1986a) Laminar and regional specificity of tangle and plaque distribution in Alzheimer's disease (AD): A quantitative study of visual and auditory cortices. Soc. Neurosci. Abstr. 12.

Lewis, D.A., Campbell, M.J. and Morrison, J.H. (1986b) As immunohistochemical characterization of somatostatin-28 and

somatostain-28(1-12) in monkey prefrontal cortex. J. Comp. Neurol. 248: 1-18.
Lindsey, J.D. and Ellisman, M.H. (1985a) The neuronal endomembrane system: I. Direct links between rough endoplasmic reticulum and the cis element of the Golgi apparatus. J. Neurosci. 5: 3111-3123.
Lindsey, J.D. and Ellisman, M.H. (1985b) The neuronal endomembrane system: II. The multiple forms of the Golgi apparatus cis element. J. Neurosci. 5: 3124-3134.
Lindsey, J.D. and Ellisman, M.H. (1985c) The neuronal endomembrane system: III. The axoplasmic reticulum: cytochemical relation to discrete elements and origin within the axon hillock. J. Neurosci. 5: 3135-3144.
Mann, D.M.A., Yates, P.O. and Marcyniuk, B. (1985) Correlation between senile plaque and neurofibrillary tangle counts in cerebral cortex and neuronal counts in cortex and subcortical structures in Alzheimer's disease. Neurosci. Lett. 56: 51-55.
Morrison, J.H., Benoit, R., Magistretti, P.J. and Bloom, F.E. (1983) Immunohistochemical distribution of pro-somatostatin-related peptides in cerebral cortex. Brain Res. 262: 344-351.
Pearson, R.C.A., Esiri, M.M., Hiorns, R.W., Wilcock, G.K. and Powell, T.P.S. (1985) Anatomical correlates of the distribution of the pathological changes in the neocortex in Alzheimer disease. Proc. Natl. Acad. Sci. U.S.A. 82: 4531-4534.
Peters, A., Palay, S. L., and Webster, H. deF. (1976) "The Fine Structure of the Nervous System," Saunders, Philadelphia.
Rogalski, A. A., and Singer, S. J. (1984a) Associations of elements of the Golgi apparatus with microtubules, J. Cell Biol. 99: 1092.
Rogalski, A. A., Bergmann, J. E., and Singer, S. J., 1984b, Effect of microtubule assembly status on the intracellular processing and surface expression of an integral protein of the plasma membrane, J. Cell Biol., 99:1101.
Rogers, J. and Morrison, J.H. (1985) Quantitative morphology and regional and laminar distributions of senile plaques in Alzheimer's disease. J. Neurosci. 5: 2801-2808.
Sternberger, N.H., Sternberger, L.A. and Ulrich, J. (1985) Aberrant neurofilament phosphorylation in Alzheimer disease. Proc. Natl. Acad. Sci. U.S.A. 82: 4274-4276.
Strehler, B.L. (1976) Time, Cells, and Aging, Academic Press, NY.
Suzuki, K. and Terry, R.D. (1967) Fine structural localization of acid phosphatase in senile plaques in Alzheimer's presenile dementia. Acta Neuropath. 8: 276-284.
Terry, R.D., Peck, A., DeTeresa, R., Schechter, R. and Horoupian, D.A. (1981) Some morphometric aspects of the brain in senile dementia of the Alzheimer type. Neurology 10: 184-192.
Wisniewski, H.M. and Terry, R.D. (1976) Neuropathology of the aging brain. In Neurobiology of Aging (ed., R.D. Terry and S. Gershon), Raven Press, NY, pp. 265-280.
Wood, J.G., Mirra, S.S., Pollock, N.J. and Binder, L.I. (1986) Neurofibrillary tangles of Alzheimer disease share antigenic determinants with the axonal microtubule-associated protein tau. Proc. Natl. Acad. Sci. U.S.A. 83: 4040-4043.
Young, S. J., Royer, S., Groves, P. M., and Kinnamon, J. C. (1987) An IBM PC-based system for three-dimensional reconstruction from serial sections, J. Electron. Microsc. Techniq., in press.

TRANSFORMATION OF THE NEUROFILAMENTOUS NETWORK COMPONENTS INTO PHF-LIKE STRANDS AND PHF PARACRYSTALS

J. Metuzals,* Y. Robitaille,** S. Houghton,***
S. Gauthier,+ and R. Leblanc**

*Department of Pathology, Faculty of Health Sciences, University of Ottawa, Ottawa, Ontario, Canada K1H 8M5
**Montreal Neurological Institute, Montreal, Quebec, Canada H3A 2B4
***Marine Biological Laboratory, Woods Hole, MA 02543
+McGill Center for the Study of Ageing, Montreal General Hospital, Montreal, Quebec, Canada H3G 1A4

The threshold theory of the causation of Alzheimer's disease (AD), as formulated by Roth et al.[1] has important implications for diagnosis, clinicopathological correlation and scientific inquiry. The theory is based on determination of the intensity of tangle formation and plaque counts carried out at the light microscope level, on one hand, and measures of dementia, on the other. Twenty years have passed since this theory was formulated. During this period of time fundamental advances have been made in three areas of importance for the theory. First, modern electron microscopic (EM) techniques provide the means to resolve the tangles into the specific and complex components of a modified neuron cytoskeleton.[2-5] Second, clinical and neuropsychological assessment of dementia has become increasingly precise and stringent in recent years. Finally, availability of frontal lobe biopsies from early Alzheimer patients provides the opportunity to define the threshold state and, thus, the beginning of the disease with greater accuracy.[6]

The present contribution summarizes new data related to all three aspects. We investigated nine early and four advanced Alzheimer patients from these three viewpoints in order to determine the irreversible threshold conditions responsible for the transformation of a normal aged person into an Alzheimer patient. An extensive presentation of the EM data is essential for this purpose. The quantification, chemical specification by immuno EM, and 3-D evaluation of the data are in progress.

We conclude from our results that the neuronal cytoskeleton undergoes specific transformations into paired helical filament-like strands which lead to the formation of the insoluble paracrystalline paired helical filament (PHF) assemblies. The neurofilamentous network (NFN) transformation plays an important role in this process, whereby segregation, posttranslational modifications and reassembly of the modified components through autocrosslinking occur.[7,8] The presence of insoluble antigens against NFN and microtubule-associated tau and MAP 2 proteins in PHF is in accordance with this conclusion.[3-5] Thus, according to our data, the threshold state is the state of irreversible segregation and posttransla-

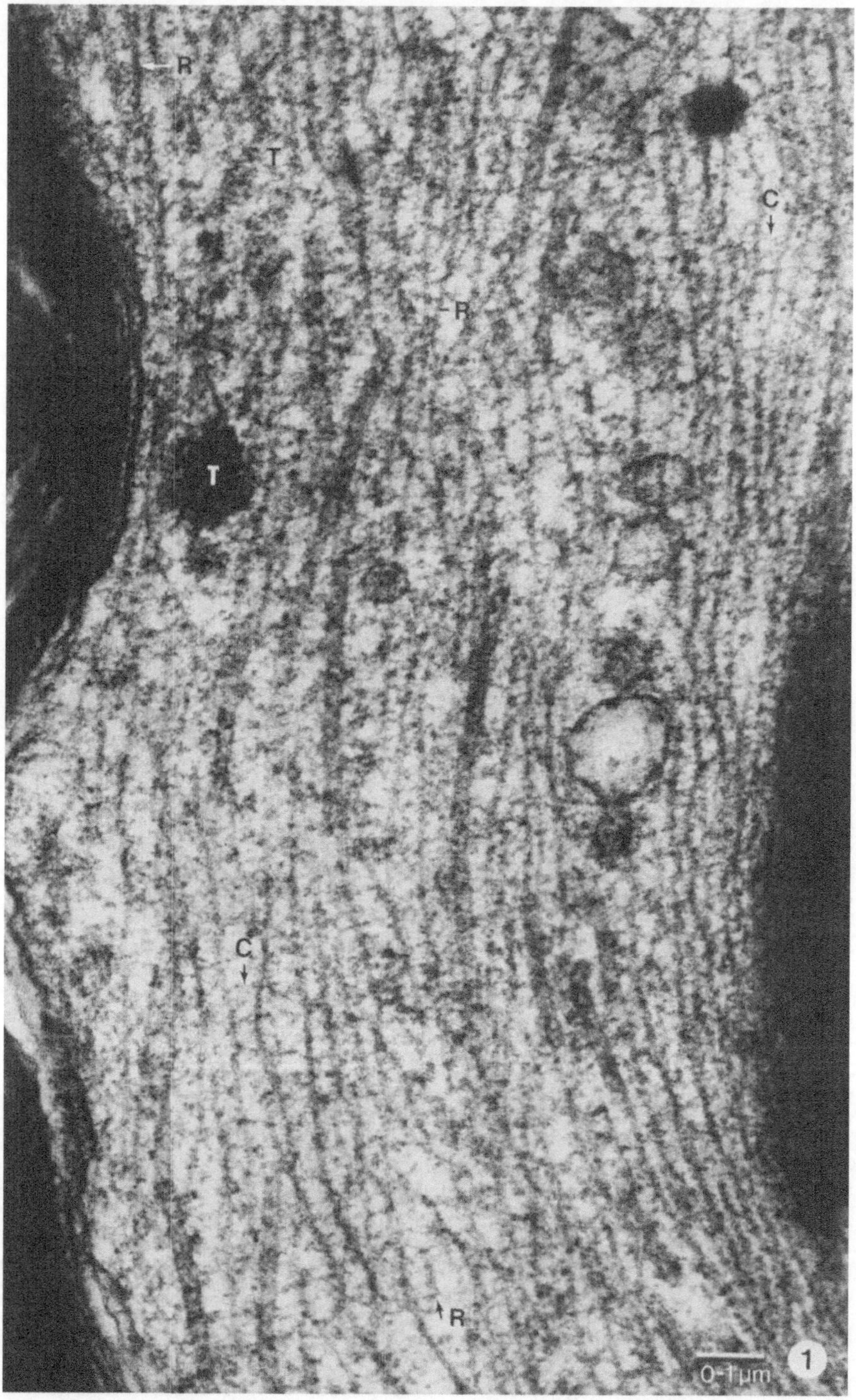

Fig. 1: Reversible segregation and rearrangement of the NFN in a normal myelinated axon: crossbridging filaments (C) and rod portion (R) of the NFN; fine filament tangle (T).

tional modifications of the NFN and the microtubule-associated protein components, as illustrated in Fig. 2. At this state a therapeutic intervention could provide hope for reversing the course of the disease.

MATERIALS AND METHODS

Frontal lobe biopsies from 13 patients with AD were investigated. The diagnosis was based on clinical, neuropsychological and histological assessment. The biopsies from 9 patients were obtained during the catheter implantation for intracerebroventricular Bethanecol infusion. The clinical features of the individual patients in the early and intermediate stages of the disease have been published in Tables 1 and 2.[6] There were four patients of the advanced NAP group consisting of 2 males and 2 females, 49-63 years old. Part of the results obtained from the EM of this group has been published.[9,10,11] All the biopsies were investigated by standard techniques of thin sectioning EM. About three thousand electron micrographs have been evaluated.

Axoplasm from 3 squid giant nerve fibers was extruded, and extracted for four hours in 0.5 ml Rubinson and Baker buffer,[12] homogenized and stored in the buffer for 48 hours. The samples were placed on grids and stained in 1% uranyl acetate, pH 7.0.

RESULTS

In electron micrographs of frontal lobe biopsies from Alzheimer patients, fixed by instantaneous immersion into the fixative, neuronal profiles can be regularly indentified as having a completely normal structure. The structure of the longitudinally-sectioned myelinated axon, illustrated in Fig. 1, is the same as that observed in myelinated axons of rabbit brain fixed by intravascular perfusion, cf. with Figs. 1,2.[9] The electron micrograph of Fig. 1 has been made from a relatively thick section to illustrate the continuous character of the cytoskeleton architecture. For a detailed description of the normal axoplasm structure, see Metuzals et al.[9] The description given for rabbit axons is also valid for normal axons of Alzheimer patients. The following conclusions can be made from this. First, the quality of human brain biopsies fixed by immersion is the same as that achieved by intravascular perfusion of rabbits under the most optimal fixation conditions. Second, this fact gives great credibility to the observations made with human biopsies and minimizes the possibility of artifactual origin of the structures analyzed.

The normal axoplasm illustrated in Fig. 1 reveals great variability in the appearance of the NFN components and an extensive array of cross-bridging filaments (C, Fig. 1). The diameter of the rod, or strand portions, of the NFN ranges from 5 nm to 30 nm (R, Fig. 1). The 2 nm protofilaments are intercoiled at different degrees determining in this way the diameter and the protofilament structure of the strand - the neurofilament (NF). In many places the NFN components are continuous with fine filament tangles of different size (T, Fig. 1). Microtubules and profiles of the smooth endoplasm reticulum and of small vesicles are interspersed among the components of the NFN.

Hypertrophied NFN revealing segregation and probably irreversible reassembly of its protein components is illustrated in Figs. 2a-c. Large profiles of neurites can be identified among normally-structured neuropil, which consist of hypertrophied NFN. Microtubules cannot be identified in these profiles but normal mitochondria are surrounded by the hypertrophied NFN. The longitudinally-sectioned NFs display a sinuous outline due to

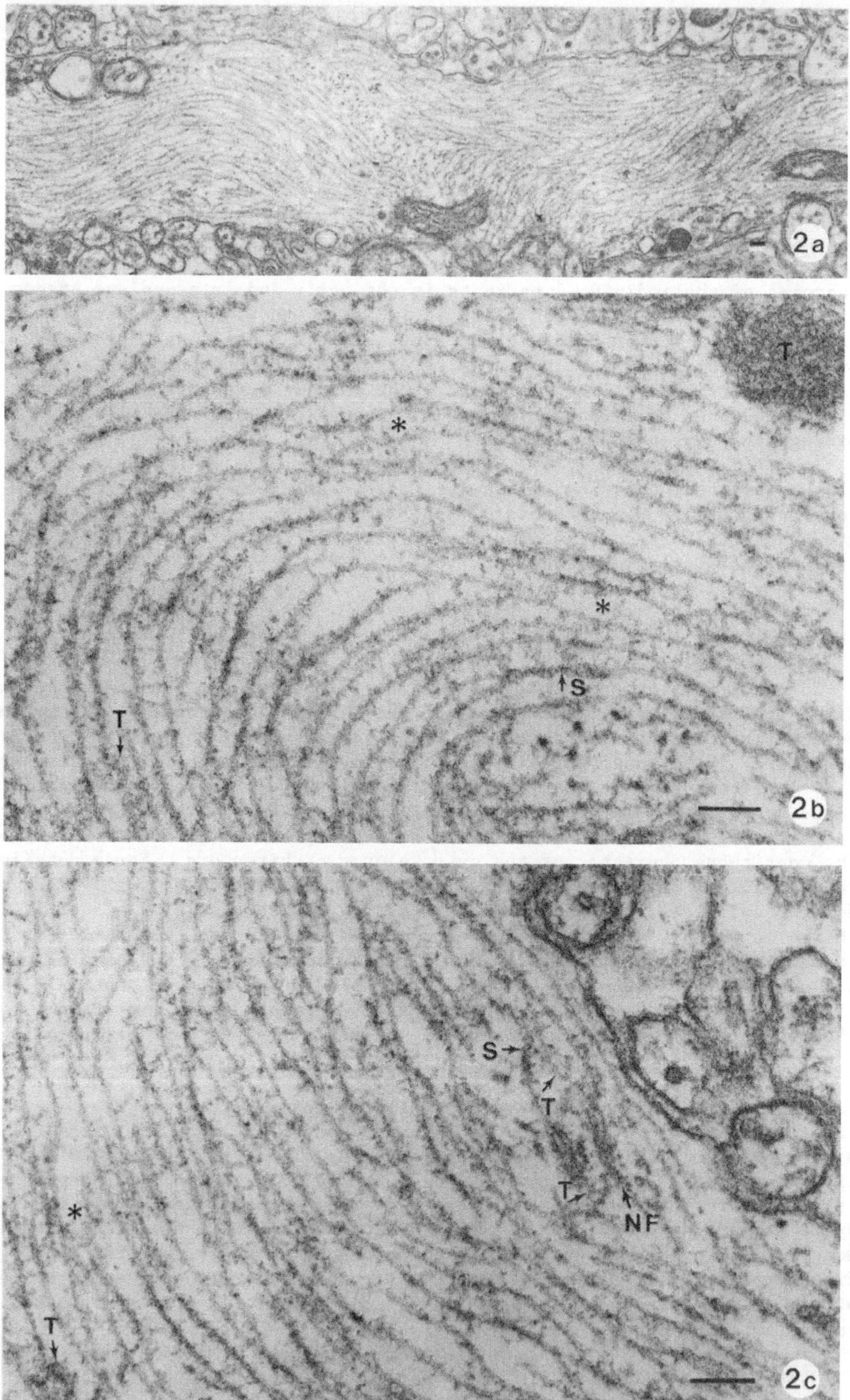

Figs. 2a,b,c: Irreversible segregation and rearrangement of the NFN in a neutrite: sinuous protofilaments (S) of a NF; lampbrush-like NF (asterisk); protofilament tangle (T); intercoiled NFs (NF). Scale bar = 0.1 μm.

their helical conformation and extensive intercoiling (Fig. 2a). Cross-sectioned and obliquely-sectioned NF profiles display a whirlpool pattern (see also Figs. 6a-d[9]). The tendency of individual NFs to twist around each other is indicated by their constant V-shaped relationships and their sinuous intercoiling patterns (NF, Fig. 2c). Sinuous protofilaments (unit-filaments) can be identified as components of individual NFs (S, Figs. 2b-c). Lampbrush-like appearances of individual NFs, due to backfolding of the sidearms (asterisks, Fig. 2b,c) as well as unravelling of individual NFs, are constant features. Such extensive hypertrophy of the NFN, lampbrush-like NF and backfolding of the crossbridges, as illustrated in Figs. 2a-c, have not yet been identified in normal rabbits. Crossbridging fine filaments are relatively less numerous as compared with normal myelinated axons, cf. Fig. 2a-c, with Fig. 1.

Extensive segregation of the 10 nm wide NFN strands into their 2 nm wide protofilaments can be observed extending over long distances in myelinated axons (Fig. 3). In these axons the protofilaments have either a parallel orientation or they are entangled. Such a segregated state of the NFN also displays the initial steps of reassembly into PHF-like structures by intercoiling of the protofilaments into irregular PHF-like assemblies and into the typical paracrystalline PHFs (F, P, Fig. 3). In other axons (Fig. 4) the amount of the PHF-like strands distributed among modified NFN is increased as compared with the amount illustrated in Fig. 3. In these axons, PHF-like strands ranging in diameter from 12 to 20 nm and displaying irregularly spaced constrictions are still continuous with NFs. Many NFs in such axons have a smooth outline and measure ~ 5 nm in diameter. The 10 nm wide NFs display irregular widenings and substructure of protofilaments that are oriented obliquely and perpendicular to the long axis of the NF.

Neuron profiles can be ordered according to the content of the NFN, fine filamentous matrix, PHF-like strands and PHFs. As the amount of the NFN and the matrix in the neuron profiles decreases the amount of PHF-like strands and PHFs increases. It can be concluded from this observation that causal relationships must exist among these assembly states of the neuron cytoskeleton in AD. The heterogeneous assembly architecture composed predominantly of densely packed PHF-like strands, ~ 15 nm in diameter and of varying periodicity of constrictions, is illustrated in Fig. 5. Interspersed among these strands are individual NFs, often in continuity (arrows, Fig. 5) with the PHF-like strands. Only some short distances of typical PHFs can be identified in such cytoskeleton assemblies as being in the state of transformation from the NFN and the matrix into typical PHF paracrystals. Structural relationships among NFN, PHF-like strands and PHF are revealed in the high resolution electron micrographs (Fig. 6), which provide support for the concept of the transformation of the NFN components into PHF. In such micrographs, large neurite profiles consisting of PHF-like strands of variable periodicity of constrictions, irregular dimensions, and variable fine structure and architecture can be identified. These strands are continuous with typical PHFs. In such neurites, NFN components are intermingled with the PHF-like strands. Continuity between components of the NFN and the PHF-like assemblies can be identified regularly (arrowhead, Fig. 6), thereby giving strong support to the transformation concept of PHF formation.[10] The great variability and irregularity in the architecture of this PHF-like assembly is illustrated in Fig. 6. The ~ 2 nm wide protofilaments and their orientation, as components of the PHF-like strands, are indicated by arrows (Fig. 6). Parallel filaments, approximately 10 nm wide, displaying railroad track structure, can be observed regularly in this assembly. Large aggregates of protofilament tangles (T, Fig. 6) are interspersed among the strands. The insert is from another region of the neurite in the same micrograph.

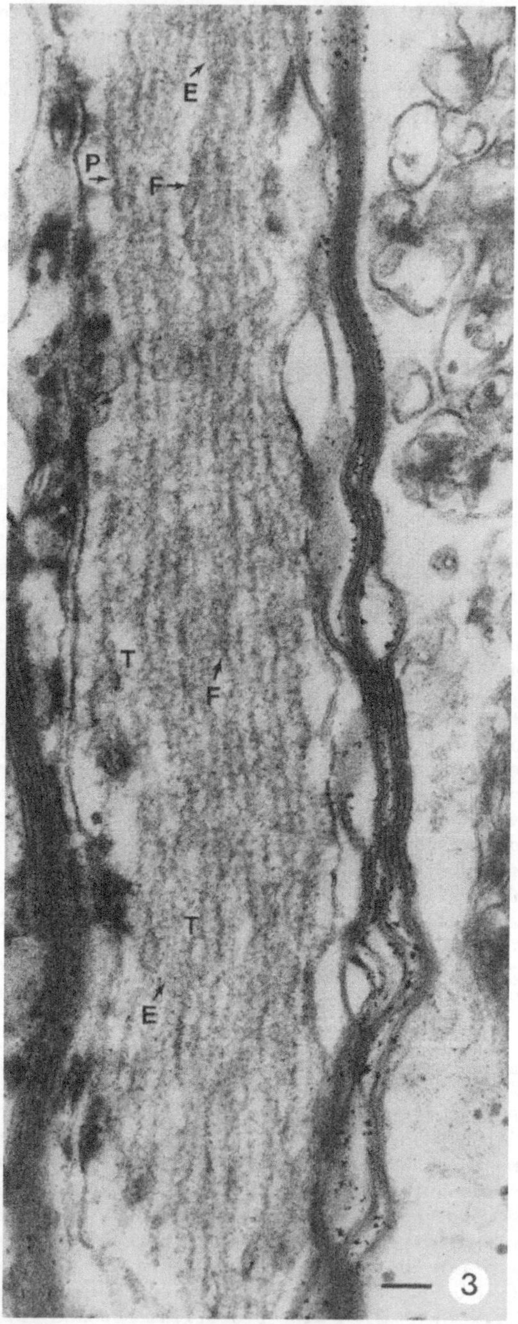

Fig. 3: Formation of PHF-like strands (F) through irreversible segregation of the NFN into their protofilaments and remodelling: protofilaments in parallel arrangement (E); protofilament tangles (T); PHF (P). Myelinated axon. Patient JDL. Scale bar = 0.1 µm.

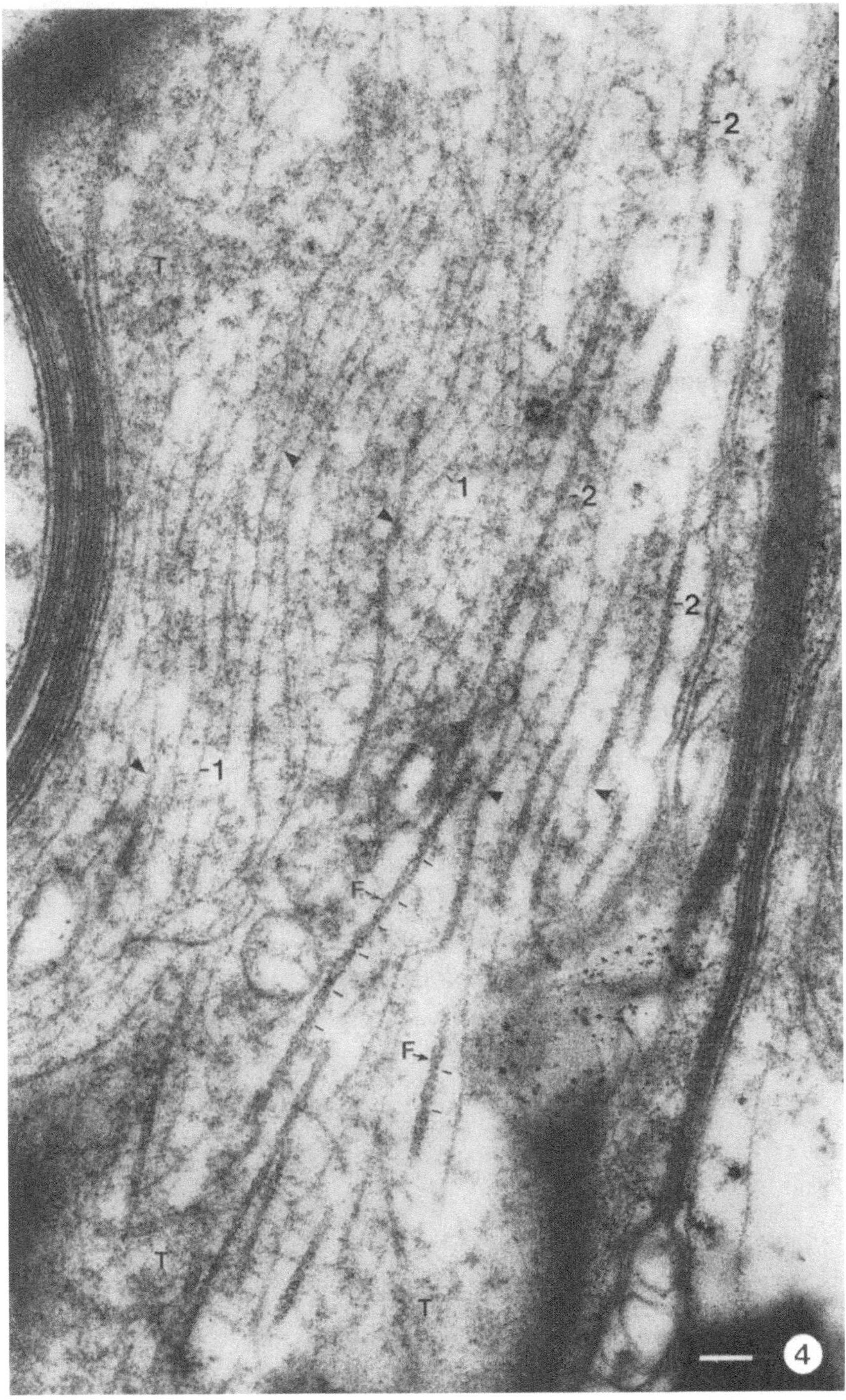

Fig. 4: PHF-like strands (F) among modified NFN: constrictions of these strands (bars); 5 nm (1) and 10 nm (2) wide NFs; continuity between NF and PHF-like strands (arrowhead); fine filament tangle (T). Myelinated axon. Patient MBR. Scale bar = 0.1 µm.

Fig. 5: The heterogeneous structure of the neurofibrillary tangle. NF (arrows) often continuous with PHF-like strands (F); typical PHFs (P). Cell body. Scale bar = 0.1 μm.

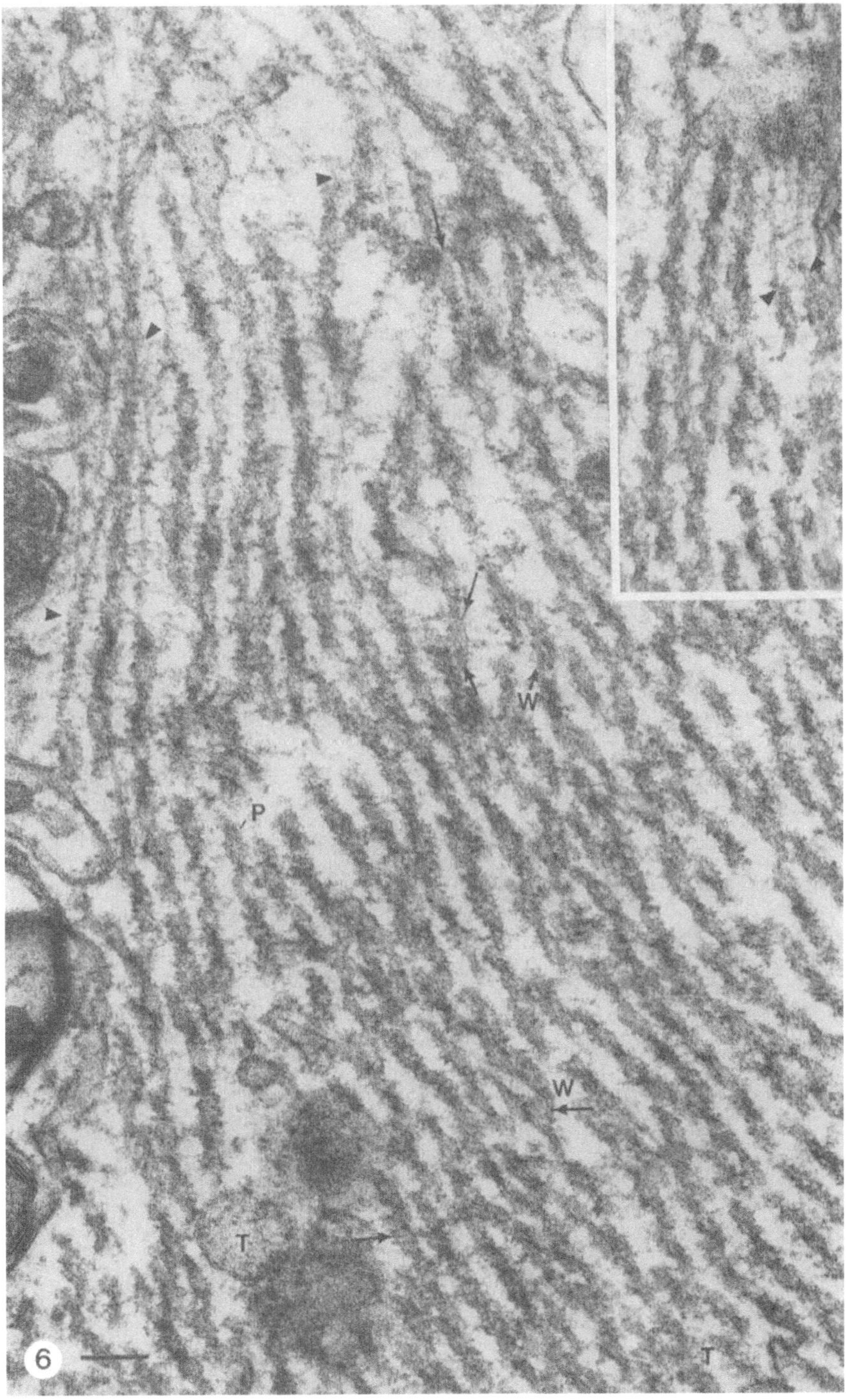

Fig. 6: Transition of the NFN into PHFs: continuity between NFN and PHF-like strand (arrowhead); protofilaments and their orientation (arrows); railroad track pattern (W); fine filament tangles (T); PHF (P). Neurite. Scale bar = 0.1 μm.

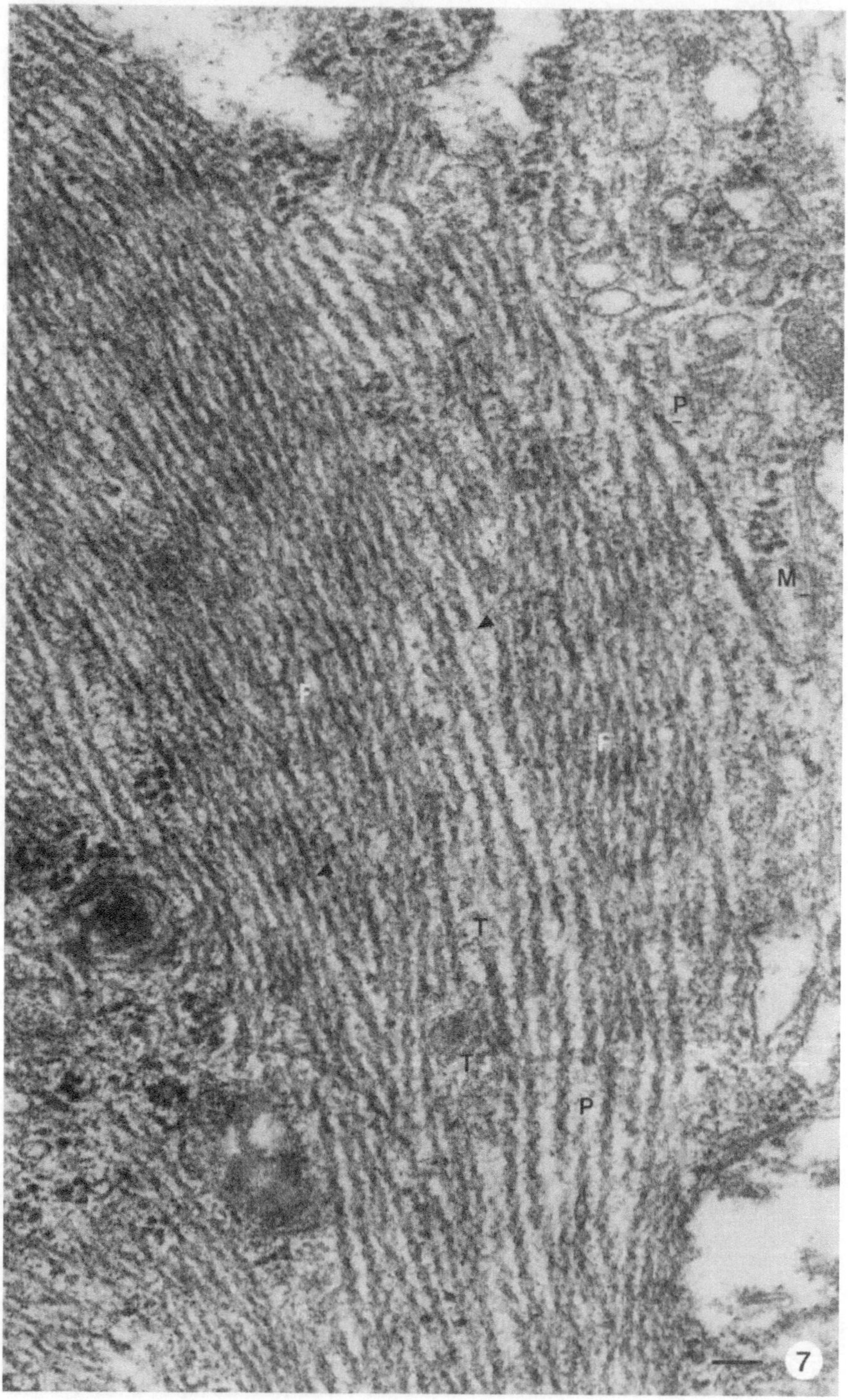

Fig. 7: Cell body. Formation of PHFs (P) by reassembly and phase transition of the fine filament tangles (T). PHF-like strands (F); continuity between PHF-like strands and PHF (arrowhead); microtubule (M). Scale bar = 0.1 μm.

The arrangements illustrated in Fig. 7 represent phase transition[9] of randomly organized and supposedly soluble protofilament tangles into the compactly ordered and insoluble state of typical PHFs. The same process is illustrated in the last figure where transition of the randomly ordered microfilament tangle is continuous with the crystalline order of a tropomyosin-like lattice.[13] An area of neurofibrillary tangle in the peripheral part of a neuron perikaryon is illustrated in Fig. 7. Intact microtubules (M, Fig. 7) can be seen in the region adjacent to the plasma membrane which is not included in the figure. Long stretches of individual, characteristic PHFs are loosely packed. Such loosely packed regions of typical PHFs are surrounded by densely packed bundles consisting predominantly of ~ 15 nm wide PHF-like strands (F, Fig. 7). Continuity between these strands and typical PHFs can be identified (arrowhead, Fig. 7). A strand ~ 50 nm wide consisting of loosely intercoiled filaments ~ 5 nm wide is located between two typical PHFs.

Formation of typical PHFs from PHF-like strands and fine filament tangles is revealed by high resolution electron micrographs taken from neurofibrillary tangles (Fig. 8). Such micrographs illustrate the detailed morphology of the dense regions of a neurofibrillary tangle and of some typical PHFs (P). The dense regions consist of densely packed ~ 15 nm wide PHF-like strands (F).[10] Fine filament tangles (T) are continuous with typical PHFs and with PHF-like strands (arrowheads). Protofilaments can be identified in the PHFs (1, Fig. 8) and in PHF-like strands (2, Fig. 8). The protofilaments appear to be tightly packed and ordered at different angles against the long axis of the PHF and the PHF-like strands.

The central pathogenic process of AD - the irreversible phase transition from fine filament tangles into typical PHFs-is revealed in extensive areas of myelinated axons (Fig. 9). The myelin sheath of such axons often does not reveal pathological changes. Considerable stretches of typical PHFs emerge from areas consisting of irregularity interwoven tangles of ~ 5 nm wide and finer protofilaments. PHF-like strands (F), 10-15 nm wide, with irregularly spaced constrictions and variable architecture, are interspersed among these structures. These structures are in the process of being investigated by 3-D EM techniques in order to determine the handedness of the helical assemblies.

In the terminal phase of the PHF formation process, paracrystalline order of PHF results, revealing constrictions with a periodicity of approximately 80 nm with a modulation of fiber diameter varying between ~ 24 nm at the widest and ~ 10 nm at the narrowest part of the filament. The ordered substructure of such PHFs is revealed in high resolution micrographs (Figs. 10,11). Protofilaments, 2 nm in diameter, are tightly packed and ordered at different angles to the long axis of the PHFs or parallel to this axis (arrows). In the segment of the PHF indicated by G in Fig. 10, protofilaments perpendicular to the long axis of the PHF, as well as small granules, can be identified similar to the structures revealed by Wisniewski and Wen.[14]

The paracrystalline properties of mature PHF assemblies are expressed in regular lattice relationships, clearly visible over large areas of cross-sectioned PHFs in cell bodies[11] and in neurites (Fig. 12a). Serial sections have been obtained from the profile consisting of the PHF lattice. In electron micrographs taken from four consecutive sections (1,2,3,4), the location of each PHF profile was marked on a transparent plate and each profile was interconnected by a wire. Fig. 12b is a photograph of the model viewed obliquely from above. The rows of PHF profiles (arrows) are rotated anticlockwise throughout the model. The connecting wires on the right side of the model display a helical twist. High resolution electron micrographs of cross-sectioned PHF profiles reveal the components of the PHF - the proto-

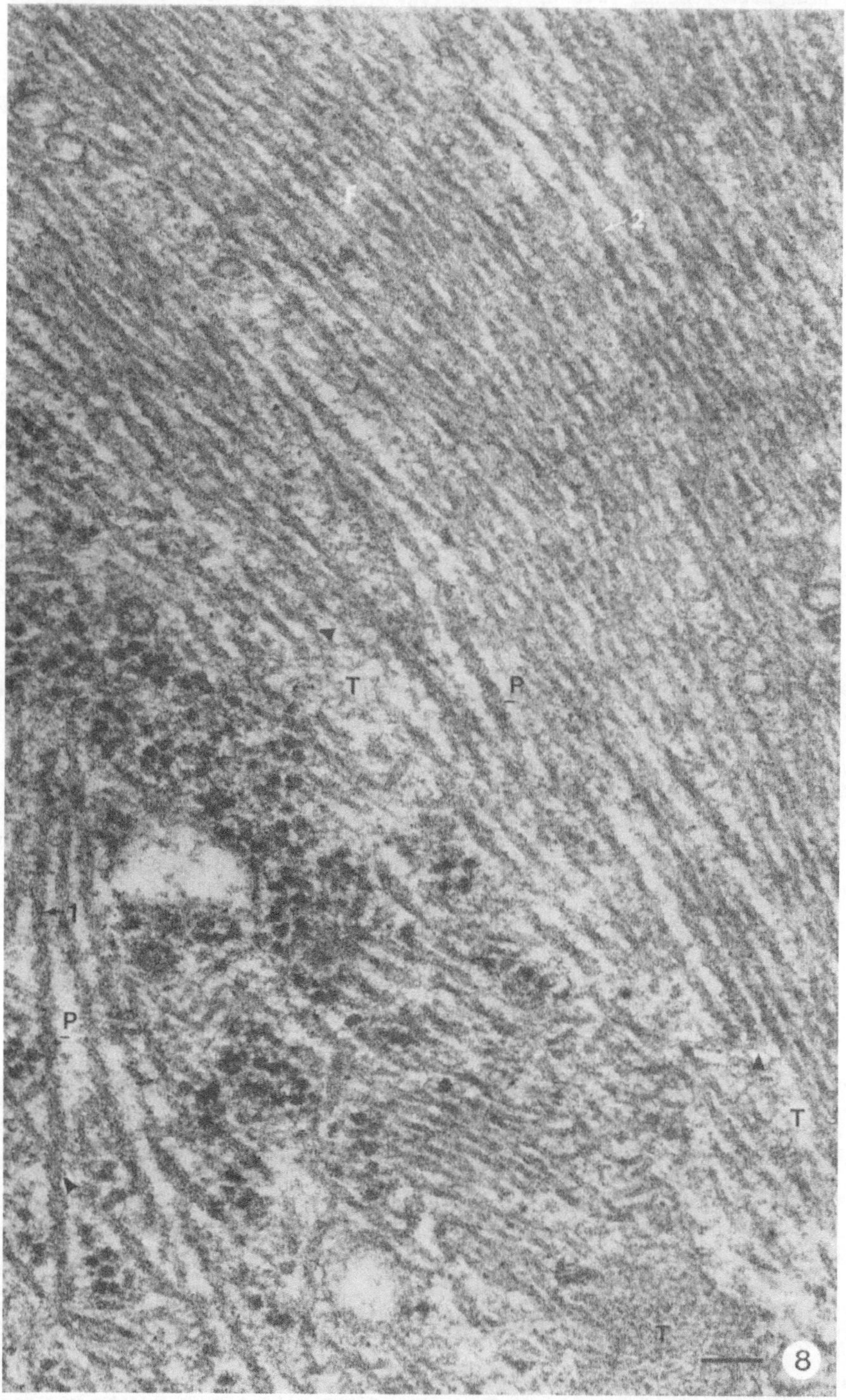

Fig. 8: Protofilaments in PHF (1) and in PHF-like strands (2). Cell body. Formation of PHFs (P) by reassembly and phase transition of the fine filament tangles (T). PHF-like strands (F); continuity between PHF-like strands and PHF (arrowhead); microtubule (M). ——— = 0.1 μm

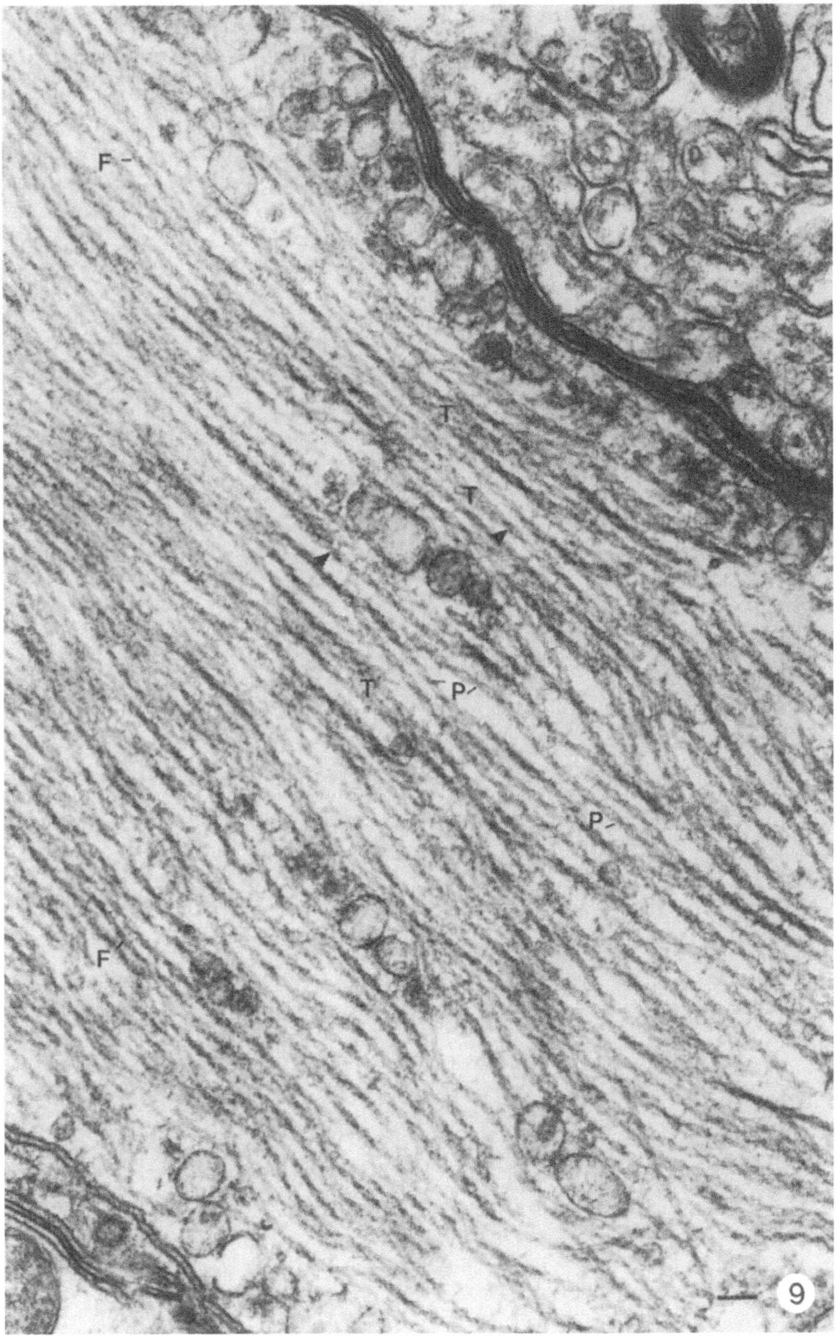

Fig. 9: Myelinated axon. Formation of PHFs (P) by reassembly and phase transition of the fine filament tangles (T). PHF-like strands (F); continuity between PHF-like strands and PHF (arrowhead); microtubule (M). Scale bar = 0.1 μm .

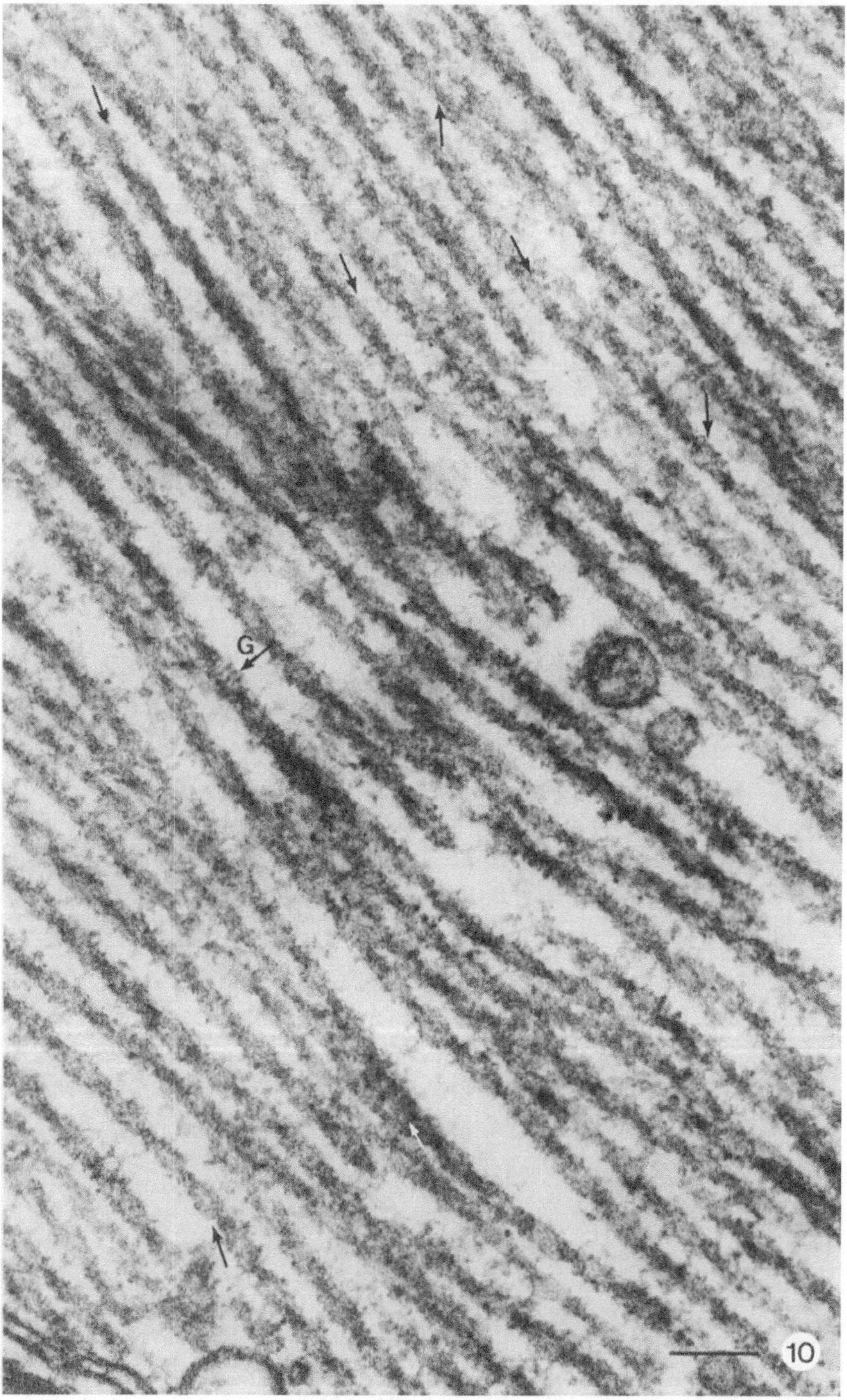

Fig 10: Cell body. Ordered filamentous substructure of PHFs: protofilaments (arrows); protofilament lattice (G). Scale bar = 0.1 μm .

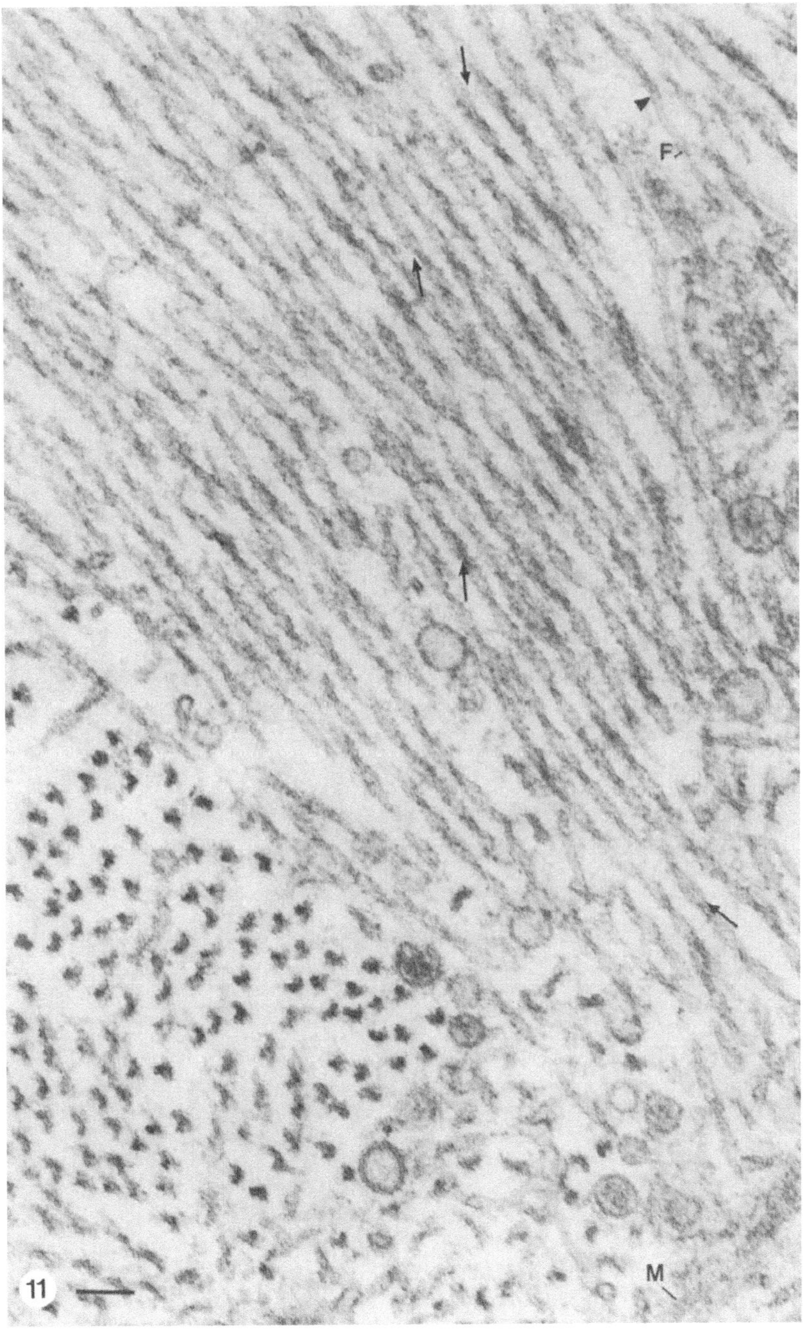

Fig. 11: Neurite. 15 nm wide strand continuous with PHF (arrowhead); microtubule (M). Ordered filamentous substructure of PHFs: protofilaments (arrows); protofilament lattice (G). Scale bar = 0.1 µm.

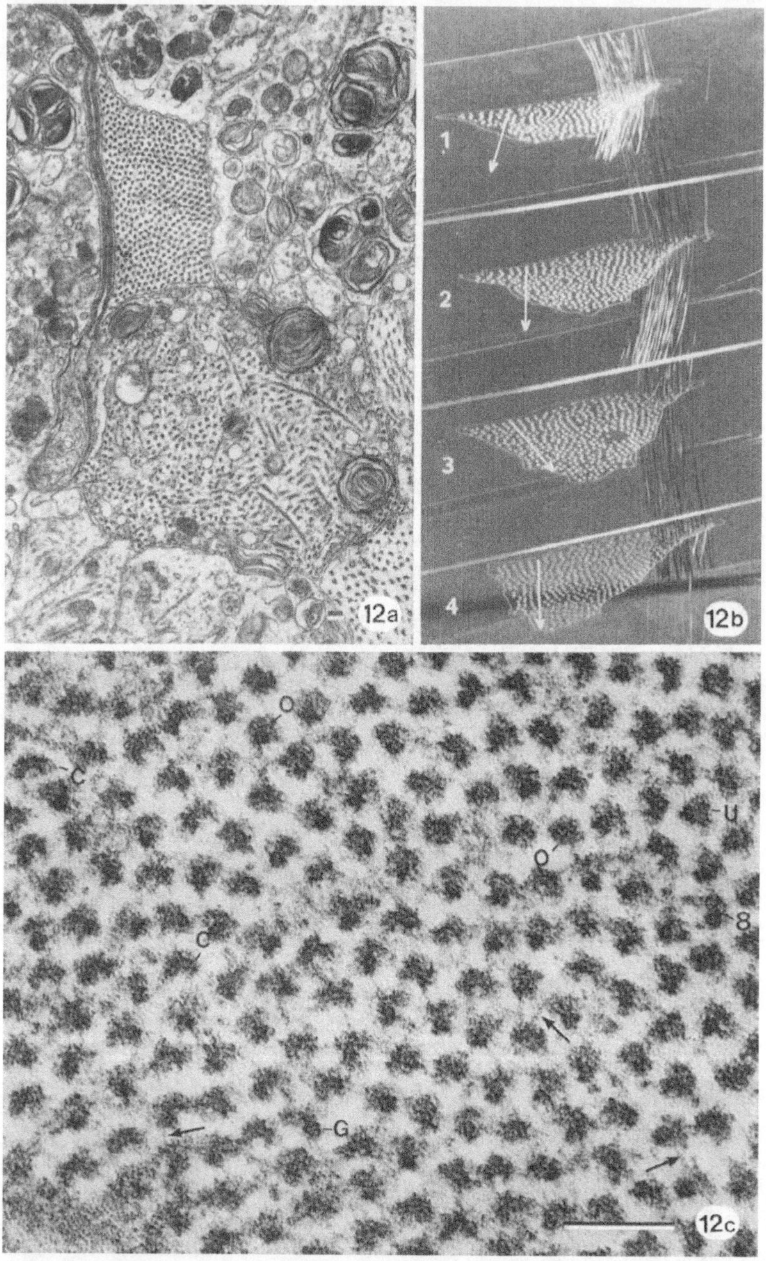

Figs. 12a,b,c: PHF lattice and substructure. b: Plastic sheets (1-4) indicating the PHF arrangement in electron micrographs taken from four consecutive sections. Arrows indicate the anticlockwise change of a PHF row. Small wires connecting the corresponding profiles demonstrate helical twisting. c: PHF structure in cross section: protofilament lattice (G), crossbridges between PHFs (arrow), PHF profiles: oval (O), crescent (C), eight-shaped (8) and triangular (U). Scale bar = 0.1 μm.

filaments - in cross section (Fig. 12c). The protofilaments appear as granules or short filaments less than 2 nm in diameter. Occasionally, patterns similar to those illustrated by the model of Wisniewski and Wen[14] can be identified (G). Crossbridges (arrows) between the filaments and the sidearm projections occur regularly. The following shapes of the PHF profiles can be identified: oval (O), crescent (C), eight-shaped (8) and triangular (R). A 3-D analysis of the PHF structure is in progress.

In vitro experiments with NF isolated fom squid giant axons and left for 48 hours in a buffer solution, support the concept derived from studies of thin sections of brain biopsies from Alzheimer patients, as presented above. NFs in these experiments intercoil in a double helical array and display morphological features similar to those observed in thin sections of Alzheimer brains.[7] Barbed wire morphology (cf. Fig. 13a with 1, Figs. 13c, d) and railroad track pattern (cf. W, Fig. 13a with W, Fig. 6) occur in both types of structures.

In electron micrographs of Hirano bodies (Fig. 14) the continuity of microfilament tangles with lattice sheets reminiscent of tropomyosin crystals is clearly illustrated.[13] Such an irreversible phase transition from a random tangle arrangement into the crystalline state may also occur during the formation of PHFs from modified components of the NFN and microtubule-associated proteins, such as tau, and MAP 2, whereby the autocrosslinking properties of these components are realized in the PHF formation process.[15]

DISCUSSION

The central question of the pathogenesis of AD is whether Alzheimer pathology is in a continuum with the normal structure, or whether the PHF represents a distinct and clear-cut break with the normal helical assemblies observed inside the neuron.[16] This break would be achieved by the de novo synthesis of a protein which assembles into the PHF.[16] In the present paper we are approaching this problem by thin sectioning of biopsies from early Alzheimer cases and by in vitro experiments. Investigations of isolated PHFs from autopsies of terminal cases are not suitable for answering this question. Our results are in accordance with the concept of Alzheimer pathology representing a continuum of modifications of the normal neuron cytoskeleton structure. This concept is in agreement with the data obtained by immuno EM of the PHFs.[3-5] Obviously, it does not exclude the possibility of participation of new gene products specific to Alzheimer's disease in tangle formation. The characteristic paracrystalline assembly of PHFs represents the terminal phase of such a continuous modification. This conclusion is based on the observations presented in Results and discussed briefly as follows.

From the analysis of the NFN in normal neurons it can be concluded that segregation and reversible reassembly of the NFN protein components into assemblies of different architecture are constantly occurring in the neuroplasm under normal conditions (Fig. 1). If the segregated components are modified posttranslationally by such processes, for example, as phosphorylation or through proteolysis by Ca^{2+} -activated proteases,[8] the modified components may reassemble irreversibly into structures of abnormal, pathological characters (Figs. 2,3,4). The dense packing of the modified filaments of the neurofibrillary tangles, their relatively smooth outline, the reduced amount of crossbridging and matrix filaments and the lack of microtubules, characterize the assembly state of the neurofibrillary tangles. From such observations the conclusion is warranted that major pathogenic events of AD are: first, stepwise modifications of the crossbridging structures of the NFN, of the fine filamentous matrix and of the microtubule-

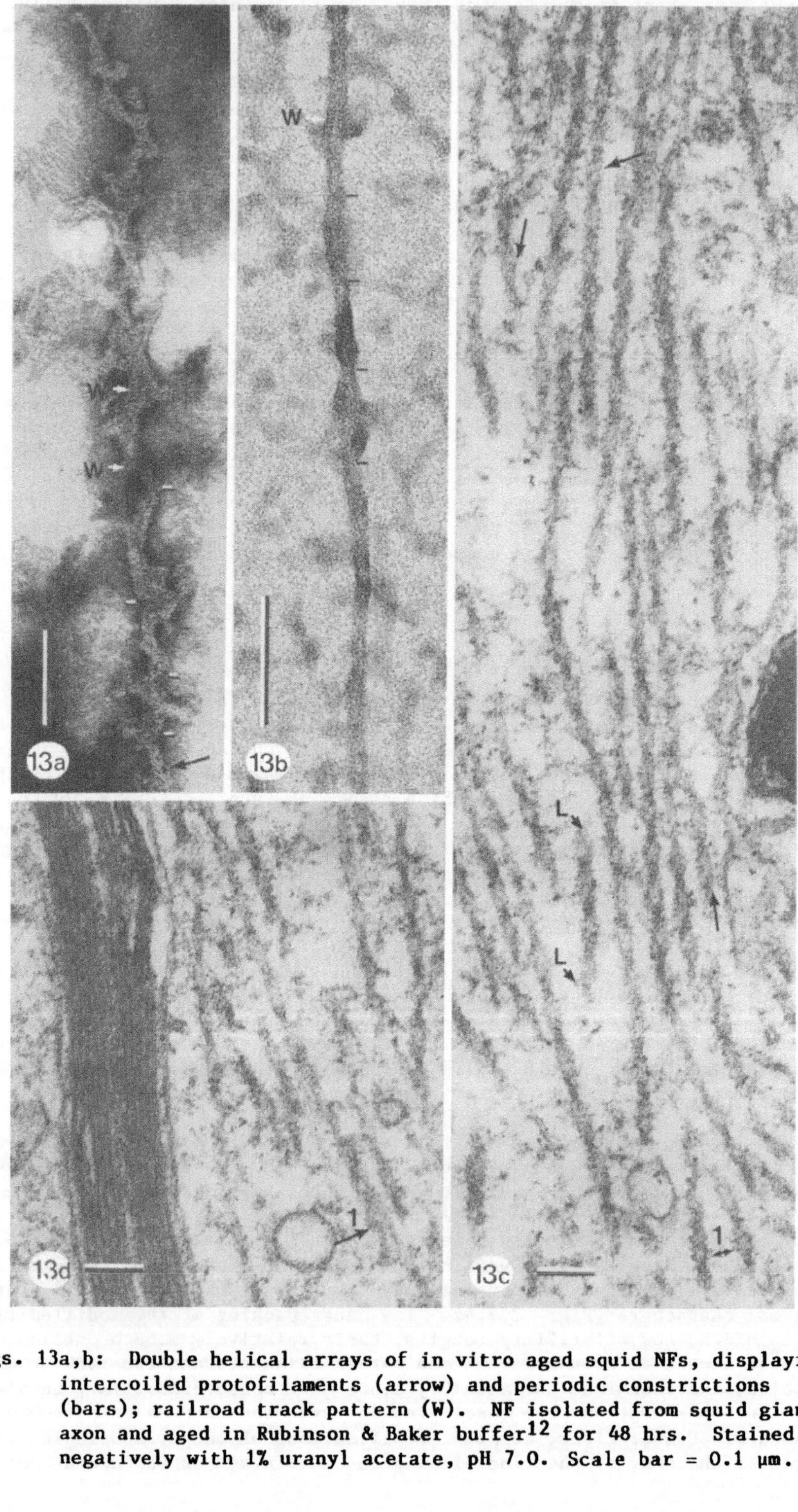

Figs. 13a,b: Double helical arrays of in vitro aged squid NFs, displaying intercoiled protofilaments (arrow) and periodic constrictions (bars); railroad track pattern (W). NF isolated from squid giant axon and aged in Rubinson & Baker buffer[12] for 48 hrs. Stained negatively with 1% uranyl acetate, pH 7.0. Scale bar = 0.1 μm.

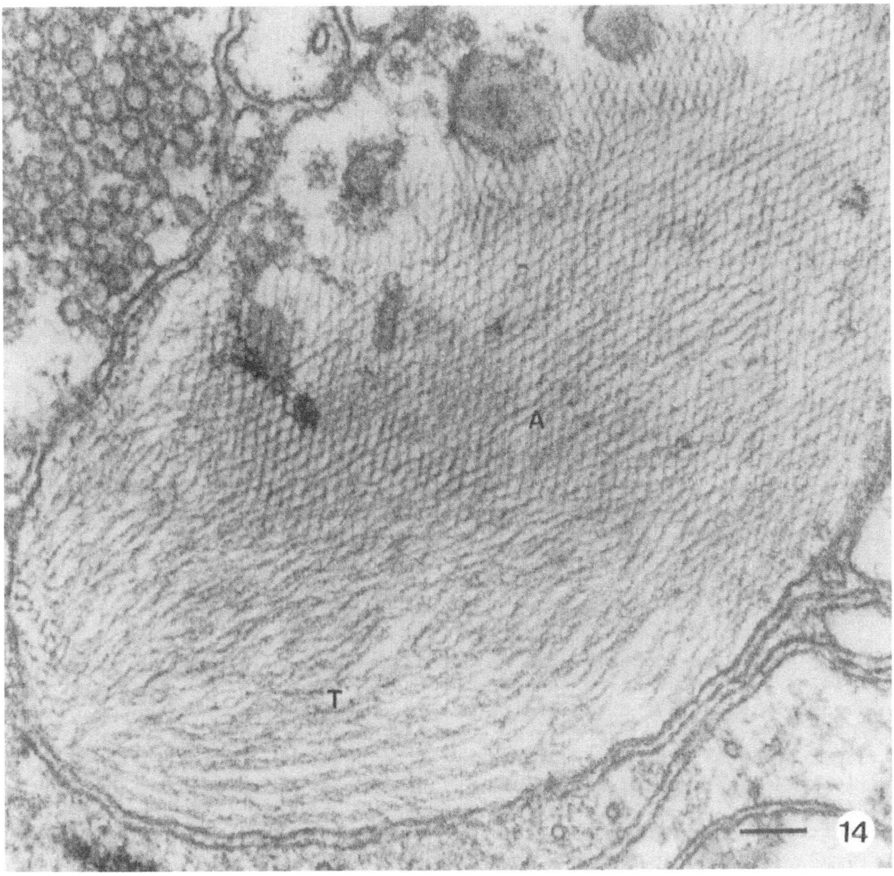

Fig 14: Phase transition of microfilament assembly in a dendrite forming a Hirano body: tangle (T) and tropomyosin-like lattice array of the microfilaments (A). Scale bar = 0.1 µm.

Figs. 13c,d: PHFs of a myelinated axon reminiscent of the double helical NF structure in Figs. 13a,b; PHFs of barbed wire morphology (1); ⟵ PHF fraying into protofilament components (L); protofilament substructure of PHFs (arrow). Scale bar = 0.1 µm.

associated proteins; second, the incorporation of the modified components into the longitudinal PHF-like strands and, third, autocrosslinking among these components (Figs. 7-10). Finally, phase transition from such an intermediary still soluble structure into the insoluble paracrystalline state of PHF is realized in this manner.

Successive increase of protofilament order occurs during the formation of PHF assembly until these filaments become ordered and stabilized in a permanent irreversible paracrystalline state. The regular lattice array is clearly visible over large areas of cross-sectioned PHFs (Figs. 12a-c).

The in vitro experiments with NFs isolated from squid giant axons support the concept derived from studies of thin sections of brain biopsies from Alzheimer patients (Figs. 13a,b). Another example of irreversible phase transition during aging and pathological modifications of the cytoskeleton is represented by the formation of Hirano bodies from random microfilament assemblies (Fig. 14). This transition of the microfilaments into the crystalline state reminiscent of tropomyosin crystals seems to be related to the chemical composition of the Hirano bodies as represented by the presence of crosslinking proteins of actin.[15] Thus, the analagous physicochemical processes - segregation, crosslinking and phase transition - are involved in the formation of the Hirano bodies as well as of the PHF assemblies.[10,13]

Conclusions can be made from early light microscopic work on neurofibrils by Cajal, Alzheimer and Bielschowsky, that are directly related to the results of this paper. Cajal[17] demonstrated experimentally the reversible temperature-dependent changes of the "neurofibrillar network" and the specific irreversible modifications induced through infection with rabies virus. The figures published by Alzheimer[18] and Bielschowsky[2] reveal that during AD chemical modifications of neurofibrils occur with a successive increase in the packing density and the degree of intercoiling of these modified filaments.

SUMMARY

Nine early and four advanced Alzheimer patients have been investigated, utilizing three approaches in order to specify the threshold state of Alzheimer's disease. Extensive thin sectioning electron microscopy of frontal lobe biopsies, correlated with stringent clinical assessment, has demonstrated that the neuronal cytoskeleton undergoes specific transformations into paired helical filament-like strands which lead to the formation of the insoluble paracrystalline paired helical filaments. The neurofilamentous network transformation plays an important role in this process, whereby segregation, posttranslational modifications and reassembly of the modified components through autocrosslinking, and phase transition occur. According to our data, the threshold state can be defined as the state of irreversible segregation and posttranslational modification of the neurofilamentous network and the microtubule-associated proteins. At this state, therapeutic intervention to reverse the disease process may be possible. The results indicate similarities between the formation of the paracrystals of the paired helical filaments and the formation of the tropomyosin-like crystals of the Hirano bodies.

ACKNOWLEDGEMENTS

This investigation was supported by grants MA-1247 and MA-8605 from the Medical Research Council of Canada to J. Metuzals. Dr. D.F. Clapin has participated in part of this investigation.

REFERENCES

1. Roth M: The association of clinical and neurological findings and its bearing on the classification and aetiology of Alzheimer's disease. Brit Med Bull 42: 42-50, 1986
2. Metuzals J (ed): Electron Microscopy and Alzheimer's Disease, San Francisco Press, Inc, San Francisco, 1986
3. Perry G, Rizzuto N, Autilio-Gambetti L, et al: Paired helical filaments from Alzheimer disease patients contain cytoskeletal components. Proc Natl Acad Sci USA 82: 3916-3920, 1985
4. Kosik KS, Joachim CL, Selkoe DJ: Microtubule-associated protein tau (τ) is a major antigenic component of paired helical filaments in Alzheimer disease. Proc Natl Acad Sci USA 83: 4044-4048, 1986
5. Perry G, Autilio-Gambetti L, Gambetti P: Neurofilament antibodies are elicited by Alzheimer paired helical filament fractions. J Cell Biol 103: 564a (Abst), 1986
6. Gauthier S, Leblanc R, Quirion R, et al: Transmitter-replacement therapy in Alzheimer's disease using intracerebroventricular infusions of receptor agonists. Can J Neurol Sci 13 (4, Suppl): 394-402, 1986
7. Metuzals J: Neurofilaments and paired helical filaments, in Metuzals J. (ed): Electron Microscopy and Alzheimer's Disease, San Francisco Press, Inc, San Francisco, 1986, p 34
8. Metuzals J, Pant H, Gainer H, et al: In vitro modifications of the neurofilamentous network isolated from the squid giant axon, in Metuzals J (ed): Electron Microscopy and Alzheimer's Disease, San Francisco Press, Inc, San Francisco, 1986, p 38
9. Metuzals J, Montpetit V, Clapin DF: Organization of the neurofilamentous network. Cell Tissue Res 214: 455-482, 1981
10. Metuzals J, Clapin DF, Montpetit V: Formation of paired helical filaments in Alzheimer's disease. Proc Annu EMSA Meet 42: 302-303, 1984
11. Metuzals J, Montpetit V, Clapin DF, et al: Arrays of paired helical filaments in Alzheimer's disease. Proc Annu EMSA Meet 40: 348-349, 1982
12. Rubinson KA, Baker PF: The flow properties of axoplasm in a defined chemical environment: influence of anions and calcium. Proc R Soc Lond B 205: 323-345, 1979
13. Metuzals J, Montpetit V, Clapin DF, et al: Filamentous arrays reminiscent of tropomyosin crystals in Hirano bodies of Alzheimer's disease. Proc Annu EMSA Meet 41: 532-533, 1983
14. Wisniewski HM, Wen GY: Substructures of paired helical filaments from Alzheimer's disease neurofibrillary tangles. Acta Neuropathol (Berl) 66: 173-176, 1985
15. Galloway PG, Perry G, Gambetti P: Hirano body filaments contain actin and actin associated proteins. J Neuropathol Exp Neurol, in press, 1987
16. Wischik CM, Crowther RA: Subunit structure of the Alzheimer tangle. Brit Med Bull 42: 51-56, 1986
17. Cajal SR: Histologie du Système Nerveux de l'Homme et des Vertébrés. Tome premier. Consejo Superior de Investigaciones Cientificas. Institute Ramon y Cajal, Madrid, 1952
18. Goldman RD, Goldman A, Jones J, et al: The structure and function of intermediate filament networks in mammalian cells, in Metuzals J (ed): Electron Microscopy and Alzheimer's Disease, San Francisco Press, Inc, San Francisco, 1986, p 60

INTRANEURONAL AND EXTRACELLULAR FILAMENTS IN THE PATHOGENESIS OF ALZHEIMER'S DISEASE

Dennis J. Selkoe

Departments of Neurology and Neuroscience
Harvard Medical School
and
Center for Neurologic Diseases
Brigham and Women's Hospital
Boston, MA 02115

STRUCTURAL CHANGES IN ALZHEIMER NEURONS: THE RATIONALE FOR MOLECULAR STUDIES

Neuritic (senile) plaques and neurofibrillary tangles (NFT) represent the principal structural alterations of neurons in presenile and senile dementia of the Alzheimer type (Alzheimer's disease). These two types of lesions are also found in much smaller numbers and in restricted topographical distribution in the brains of many neurologically normal individuals after the seventh decade. Tangles, plaques and other structural changes found in Alzheimer brain tissue (amyloid angiopathy of cerebral and meningeal vessels, granulovacuolar degeneration of hippocampal neurons, lipofuscin deposition in neurons and glia) are, therefore, qualitatively indistinguishable from the lesions that accompany normal brain aging but are quantitatively much increased in AD. Neurofibrillary tangles (NFT) are large, non-membrane-bound cytoplasmic masses of abnormal fibers found within the perikarya of neurons in hippocampus, amygdala, cerebral cortex and certain deep gray nuclear structures. Numerous electron microscopic studies have concluded that the vast majority of fibers in these tangles have the appearance of tightly adherent pairs of helically wound, intermediate-sized (10 nm) filaments, referred to as paired helical filaments (PHF), and having a maximum width of about 20 nm and a half-periodicity of roughly 80 nm (1,2,3). Many of the abnormal neurites (both axonal terminals and dendrites) making up the periphery of the senile plaque contain PHF (4). Several laboratories have reported that bundles of straight (non-helical) filaments of approximately 15-20 nm width may also be found in neurons in AD brains and may coexist with PHF in the same cells (5-8).

The accumulation of abnormal protein fibers (i.e., PHF and straight filaments) in neuronal cell bodies and in neuritic plaques has been statistically correlated with the principal features of AD, including degree of dementia (9,10), extent of neuronal loss in hippocampus (11) and degree of deficiency of cortical cholinergic function (12,13). Shrinkage and loss of large cholinergic neurons in the basal forebrain in AD is associated with the presence of NFT in many remaining neurons. For example, we have recently used PHF-specific polyclonal antibodies and acetylcholine-esterase histochemistry in double-labeling experiments that show that (a) loss of cholinergic neurons in the nucleus basalis of Meynert (nbM) is associated with extensive NFT formation, including many residual NFT ghosts or tombstones in loci where nbM neurons once existed; and (b) numerous non-cholinergic neurons and neurites in basal forebrain show NFT and plaque formation (14). Importantly, those cases of AD with a clearly defined genetic basis (autosomal dominant AD) show the most severe and extensive neurofibrillary lesions, affecting innumerable neurons and neurites throughout the brain. Furthermore, older patients with Down's syndrome often show marked NFT and neuritic plaque formation.

In addition to the intraneuronal fibrous changes just described, AD brain tissue displays abundant extracellular deposits of abnormal filaments that are found in the cores of senile plaques, in small bundles in the cortical neuropil and in the walls of certain leptomeningeal and cortical blood vessels (Congophilic angiopathy). Like the NFT, these fibrous extracellular deposits show characteristic green birefringence under polarized light after staining with the dye Congo red and have thus been classified as amyloid deposits. EM demonstrates that the filaments comprising these deposits are generally in the range of 6 to 10 nm in diameter (mean ~8 nm) and are unpaired. Their fine structure is distinct from that of the intraneuronal PHF and straight filaments. The protein chemical relationship of the intraneuronal PHF and extracellular amyloid filaments that accumulate in AD is a subject of active investigation.

Thus, in contrast to Huntington's disease, in which striatal neuronal loss is unaccompanied by any characteristic ultrastructural change in adjacent neurons that have not yet died, neuronal attrition in Alzheimer brain is associated with marked structural changes in many remaining cells and in the neuropil. In view of these observations, it is important to identify the molecular nature and origin of PHF and amyloid fibers. The principal objective of current work is to characterize the molecular mechanism of amyloid fiber and PHF formation and establish what role these processes may play in accelerated death of neurons during normal aging and in Alzheimer's disease. In setting this goal, one must emphasize that neurofibrillary tangle formation is a somewhat non-specific event occurring not only in normal aging but in a number of etiologically diverse human brain diseases. On the other hand, neuritic plaques with their central amyloid cores are a much more specific marker for Alzheimer's disease than is the neurofibrillary tangle. Senile plaque formation is largely restricted to AD, normal brain aging and the AD-type brain degeneration that occurs in virtually all patients with Down's syndrome after the third or fourth decade. Therefore, understanding the mechanism of neuritic plaque formation is likely to provide more direct clues to early events in Alzheimer's disease than are studies of NFT constituent PHF.

Biochemistry of the Neurofibrillary Tangle

Most early studies directed at identifying the constituent proteins of PHF and other fibers in the NFT employed immunocytochemistry on *in situ* tangles and senile plaque neurites at the light microscopic level, using a variety of polyclonal or monoclonal antibodies (reviewed in reference 15). In general, these studies showed that numerous polyclonal and monoclonal antibodies against neurofilaments (the normal 10 nm protein filaments found in virtually all neurons) do not cross-react with NFT and senile plaque neurites, whereas certain neurofilament antibodies do. Certain antibodies against microtubule-associated proteins but not antibodies to tubulin itself (the major subunit protein of microtubules) have also been shown to label NFT (16). The results of early immunocytochemical studies suggested a complex antigenic composition of the abnormal fibers, including portions of neurofilament and microtubule-associated proteins. Since such studies, even using monoclonal antibodies at the EM level, seemed unlikely to provide comprehensive data about the protein composition and thus the molecular origin of PHF, our laboratory carried out experiments aimed at characterizing PHF and their bonding structure by the techniques of analytical protein chemistry.

We initially used extraction with the non-ionic detergent Triton X-100 and subcellular fractionation, as conventionally employed in cytoskeletal studies, to derive partially enriched PHF fractions from AD cortex. One- and two-dimensional gel electrophoresis failed to demonstrate any consistently increased or unique neuronal polypeptides in PHF-enriched fractions compared to similar fractions from control brain (17). After repeatedly finding no consistent change in identifiable neuronal proteins in PHF-enriched fractions, we postulated that PHF might consist in part of insoluble proteins that did not enter gels after heating in sample buffer containing sodium dodecyl sulfate (SDS) and reagents that reduce disulfide bonds [e.g., β-mercaptoethanol (βME)]. Electron microscopy of the SDS/βME-insoluble residue of our PHF preparations excluded at the top of the stacking gel demonstrated abundant PHF which retained their 80 nm periodicity but were slightly thinner (~16-20 nm) in diameter than in situ PHF (17). We subsequently developed a method for isolating intact NFT and found that many, though not all, NFT appeared to remain intact and retained their characteristic Congo red birefringence after heating in SDS/βME buffer (17). Further analyses showed that many PHF remained insoluble and recognizable by electron microscopy after treatment with several strong denaturants, including 9.5 urea, 6 M guanidine HCl and phenol (17,18). The straight 15 nm filaments found in degenerating neurons in some cases of AD can also be recovered in the SDS/βME-insoluble fraction of cortex. Since no quantitative assay for PHF in intact cortex or aqueous samples was available, we could not estimate the percent recovery of PHF after SDS/βME extraction (17). It is probable that a portion of the PHF fiber and/or some whole fibers are solubilized by harsh solvent extraction. However, the marked enrichment of PHF after SDS/βME treatment, the preservation of their characteristic helical periodicity, the fact that actual manual counts of NFT before and after treatment indicate no quantitative loss (17) and the absence of soluble polypeptides released into SDS/acrylamide gels all suggest that most PHF fibers are highly stable protein polymers. At the same time, it is clear from the studies by Dr. Iqbal and colleagues that a percentage of NFT in some cases of AD are disaggregated by heating in SDS (19).

The discovery of the high degree of inertness of the fibers comprising NFT led to several conclusions. First, PHF are difficult to analyze by conventional biochemical techniques. Second, their insolubility can be used to advantage to purify them partially by extraction with harsh solvents. Third, and most important, their apparent high degree of chemical inertness is suggestive of a protein polymer containing multiple strong noncovalent bonds, and/or involving some covalent crosslinking.

The shared properties of PHF fibers and known crosslinked protein polymers led us to hypothesize that covalent intermolecular crosslinking might be involved in PHF formation. A number of mammalian intracellular proteins have been shown to be crosslinked by glutamine-lysine bonds formed by the calcium-activated enzyme, transglutaminase. We have demonstrated that NF and other endogenous CNS proteins can serve as substrates for brain transglutaminase in vitro (20). However, it has not been possible to demonstrate directly the presence of such putative crosslinks in PHF since their identification requires the sequential degradation of the protein by proteolytic enzymes and subsequent isolation of the dipeptide crosslinks, a technique which cannot yet be carried out on the highly inert PHF. Furthermore, PHF have not, to our knowledge, been purified to homogeneity, allowing such an assay. Although the possibility of covalent crosslinking in PHF remains, an alternative mechanism for explaining the insolubility of PHF derives from recent x-ray diffraction studies.

We have carried out x-ray diffraction analysis of isolated PHF and senile plaque amyloid cores (21). Using considerably enriched but not fully purified PHF preparations, we observed characteristic ~4.76 A and ~10.6A reflections consistent with a macromolecule containing cross-β pleated sheet conformation. This finding appears to confirm the long held hypothesis that, like other amyloids, the intraneuronal PHF in AD contain β-pleated sheet structure. Analyses on highly purified senile plaque core amyloid gave similar patterns (21). The apparent β-pleated sheet conformation of PHF and amyloid fibers could explain, at least in part, their high degree of insolubility. However, the lack of quantitative solubilization of PHF and amyloid cores by concentrated guanidine HCl differs from the behavior of amyloid fibers in non-neural organs and suggests an even greater degree of insolubility in the fibers of AD.

Although considerable enrichment of PHF has resulted from the use of strong detergent during purification (17,19,22-25), other SDS/βME-insoluble constituents of AD cortex (e.g., lipofuscin granules, amyloid fibers and unidentified granular material) still co-purify with PHF. Because of their unusual nature, it has not been possible to obtain PHF that are at or close to structural homogeneity using biochemical purification methods alone. Consequently, PHF-specific polyclonal antibodies (22,24) and monoclonal antibodies (26-28) have been produced to use as specific ligands for the further immunochemical characterization of PHF. Analysis of our polyclonal PHF antibodies (22) demonstrated that they reacted at high dilution with both in situ and SDS-extracted NFT and senile plaque neurites but, unexpectedly, they labeled no structures in normal aged human or animal brain, other than the occasional senile plaque neurites and rare NFT found in normal aged human cortex. These PHF antibodies also failed to react with several cytoskeletal proteins in human brain homogenates or brain filament fractions (e.g., NF triplet polypeptides, tubulin, actin, GFA) by Western blotting.

Recent work in several laboratories (29-33), has demonstrated that PHF antibodies show strong cross-reaction with the microtubule-associated phophoproteins designated tau (a family of closely related proteins of MW 50-65 kD). Even PHF extensively extracted in SDS remained immunoreactive with antibodies to native tau (32) suggesting that the tau antigens are an integral component of the fibers. The fact that previous Western blotting of aged normal brain homogenates with PHF antibodies did not show clear tau bands (22) may be due to the low concentration or lack of postmortem stability of tau in aged human postmortem brain. The recent studies that found that tau proteins were specifically labeled by PHF antibodies used highly-enriched fractions of microtubule-associated proteins purified from fetal brain. It appears that the tau protein that is strongly associated with or incorporated into PHF in AD is altered from normal tau. This conclusion is based on the fact that the tau associated with PHF is resistant to formalin fixation, is insoluble (compared to the high degree of solubility of native tau protein in aqueous buffers), may be abnormally phosphorylated and is partially resistant to digestion with proteases such as proteinase K (see ref. 34).

The identification of tau as an antigenic constituent of PHF raises the question of precisely how tau is altered in the disease process. Furthermore, cDNAs encoding both mouse (35) and human tau (36) and human MAP2 (37) have been cloned, enabling studies of the level of expression of tau and MAP2 in AD versus normal neurons using in situ hybridization and Northern analysis. In addition to the tau protein, specific epitopes of the microtubule-associated protein designated MAP2 shows some antigenic relationship to the PHF (16). Moreover, PHF polyclonal antisera contain antibodies other than those directed against tau or MAP2, indicating that there are other as-yet-unidentified protein components contributing to the PHF. For example, extensive absorption of a PHF antiserum with tau to the point that tau proteins on Western blots are no longer recognized still leaves the antiserum capable of immunostaining tangles and neurites in AD brain sections. One of the non-tau constituents of PHF has now been identified as ubiquitin by the elegant studies of Mori et al. (38). At this juncture, it appears that the PHF is a complex, possibly heteropolymeric fiber that derives, at least in part, from altered forms of normal neuronal microtubule-associated proteins, particuarly tau. The previous immunocytochemical evidence that neurofilaments contribute antigens to PHF must now be reinterpreted in view of recent data (39) demonstrating that several monoclonal neurofilament antibodies that are directed against phosphorylated sites on NF and also recognize PHF show cross-reaction with phosphorylated forms of tau. It appears that these monoclonal neurofilament antibodies recognize similar phosphorylated epitopes on both NF and tau. Since the vast majority of NF antibodies do not recognize tangles whereas virtually all tau antibodies tested to date do, it is likely that the previous reported reaction of certain NF antibodies with PHF is due to their cross-reaction with tau. Independent evidence for the presence in the PHF of NF epitopes not shared with tau will now be required.

Purification and Characterization of Senile Plaque Amyloid: Comparison to PHF

In order to separate senile plaque amyloid from other constituents of AD cortex and characterize it, we decided to use a novel strategy, fluorescence activated cell sorting, to isolate intact amyloid cores in an aqueous suspension (40). The discrete, roughly spherical nature of the cores and the fact that their size (6-20 microns in maximal diameter) is similar to blood cells that are routinely separated by flow cytometry led us to attempt this approach. Initial core fractions were prepared by centrifuging SDS-extracted homogenates of AD cortex at low speed, allowing quantitative pelleting of the cores while many lipofuscin (LF) granules, NFT and small particles remained in the supernatant. The suspension was then sieved and separated on sucrose density gradients. Gradient interfaces that were particularly enriched in cores were subjected to fluorescence activated cell sorting (FACS). Using fluorescence and light-scattering (size) discriminants, unlabeled cores were separated from most autofluorescent LF granules by an initial FACS sort. The cores were then rhodamine-labeled with an antiserum raised against partially purified amyloid cores and were separated from remaining contaminants by a second FACS sort. Brightfield fluorescence and polarization microscopy of the final fraction revealed highly purified, intact amyloid cores. EM showed that the cores were virtually free of contaminants except for occasional aggregates of LF. By particle counting, the final fractions consisted of ≥ 90% amyloid cores, the remainder being LF and rare capillary fragments.

We carried out amino acid analyses of purified amyloid cores (40). Each plaque core contains approximately 60-130 pg of protein. The amino acid composition shows approximately 50% non-polar resideues with glycine being the most common residue (~18%). These analyses showed marked similarity to the amino acid composition published by Glenner and Wong (41) for a meningeal vascular amyloid peptide (the β-protein) prepared from AD brain and not found in normal aged meningeal vessels.

In contrast to the meningeal vascular amyloid protein characterized by Glenner and Wong (41) the amyloid cores are not quantitatively depolymerized in guanidine hydrochloride. However, following treatment in the stronger salt, guanidine thiocyanate (6.8 M), birefringent cores are no longer seen and amyloid filaments are not visible by EM. Similar effects occur when the purified cores are treated with 88% formic acid at room temperature for 1 hour (40,42). Heating in SDS sample buffer and electrophoresis of such formic acid or guanidine SCN-treated cores on urea/acrylamide gels (Coomassie blue staining) reveals: (1) a diffuse smear of Coomassie-positive material throughout the entire running gel; (2) accentuation of this smear at the lower end of the gel, forming two broad bands at approximately 4-7,000 and 11-15,000 MW; and (3) some excluded, insoluble material remaining at the top of the running and stacking gels (40). We have subsequently treated partially purified PHF fractions with either 80% formic acid or 7 M guanidine SCN followed by treatment with 1% SDS and have observed similar broad low MW bands which co-migrate with the bands released from the amyloid cores (40). These bands are not present in control brain fractions free of PHF or cores.

The low molecular weight proteins released from amyloid cores or from PHF fractions have been separated by HPLC and then used as immunogens to raise polyclonal antibodies. The latter show strong reaction with senile plaque cores and vascular amyloid deposits in AD brain, but no reaction with NFT or the neurites of senile plaques. Various absorption experiments confirm the specificity of these antibodies for amyloid fibers, not PHF. It appears, therefore, that the amyloid fibers are principally composed of a low molecular weight, hydrophobic protein that readily aggregates to form polymers. The presence of similar proteins in PHF-enriched fractions may be due to contaminating amyloid filaments in these fractions rather than to the PHF. Virtually all available immunochemical evidence from numerous laboratories suggests that the amyloid filaments found extracellularly in AD brain tissue are not composed of the same intraneuronal proteins (including tau and ubiquitin) that comprise the PHF. Furthermore, PHF occur abundantly in several human neurologic diseases in which one sees no amyloid deposition or neuritic plaque formation whatsoever. Also, amyloid filaments that are immunochemically and protein chemically highly similar to those in senile plaque cores occur in the walls of meningeal arteries outside of the brain, a location that makes a direct neuronal origin less likely. However, further careful work will be needed to determine whether the PHF contain a central protein filament that is composed of the same β-polypeptide that the extraneuronal amyloid filaments are; such a possibility cannot be excluded at present.

Whereas Glenner and Wong were able to obtain a protein sequence up to residue 28 for the meningeal vascular amyloid peptide, attempts in our laboratory (40) and others (43; D. Miller and H. Wisniewski, personal communication) to sequence the core-derived low M_r protein have been unsuccessful. The protein behaves as if it has a blocked n-terminus, in contrast to the meningeal vascular β-protein. In contrast, we have been able to purify the meningeal vascular amyloid protein either according to the method of Glenner and Wong or using an SDS-method modified from our core preparation and find that it is readily sequenced and shows the same sequence as reported by Glenner and Wong to residue 28 (Joachim, Duffy and Selkoe, unpublished data). Further attempts to obtain unambiguous protein sequence beyond residue 28 are under way. The results just summarized indicate that although the same β-protein may be present in vascular amyloid cores in AD brain, the latter deposits have a modified form of the protein in that it is insoluble in guanidine hydrochloride and appears to have a blocked n-terminus.

Conservation of AD-Type Amyloid Proteins in Other Aged Mammals

We have prepared a panel of polyclonal antibodies to intact amyloid filaments, their low molecular weight β-protein, or a synthetic amyloid β-peptide and have used these to examine amyloid deposits in aged monkeys and other species (44). It was reported some years ago by Wisniewski and colleagues (45,46) that aged monkeys and dogs sometimes display senile plaques. Furthermore, Struble and co-workers (47,48) have demonstrated that the neurites comprising these monkey plaques have varying transmitter specificities, just as is the case for AD neuritic plaques. We have now shown that both the amyloid cores of such monkey senile plaques as well as microvascular amyloid deposits in meningeal and cortical vessels are immunochemically highly similar to those in AD brain (44). Absorption of the antisera with their respective human amyloid antigens abolishes the staining of both human and animal plaque and vascular amyloid. These results indicate that the amyloid proteins that are deposited progressively in AD and to a much lesser extent in normal aged human brain are also formed into insoluble amyloid filaments in a number of aged mammals, including squirrel monkey, rhesus monkey, orangutan, dog and polar bear (44). As a result, any hypothesis about the genetic and post-translational mechanism underlying amyloid deposition in AD and Down's syndrome will need to take into account the apparently well-conserved process of amyloid deposition in normal aged mammals, including humans.

Amyloid Antibodies Provide Evidence for Amyloid-associated Proteins Distinct from the β-proteins

In characterizing several polyclonal antibodies to AD amyloid developed in our lab, we observed that certain of these antisera had very low or no reactivity with the amyloid β-protein. As noted above, the β-protein appears to be the principal constituent of amyloid filaments in both cerebral vessels and senile plaque cores. However, polyclonal antibodies raised against fractions of intact amyloid filaments (i.e., not solubilized) recognize the β-protein weakly but label amyloid deposits in plaque cores and vessels at very high titers. We have, therefore, tentatively categorized our amyloid antibodies into three groups, according to the immunogens used: (1) Antisera to intact amyloid filaments; (2) antisera to the low molecular weight hydrophobic protein (β-protein) solubilized by formic acid from such amyloid filaments; and (3) antibodies to synthetic β-protein peptides. Groups two and three show reaction with the β-protein by dot immunoassay and on Western blots; in contrast, group one antisera show weak reaction. Nonetheless, all three groups of antisera stain amyloid deposits in AD brain sections in a highly similar, if not identical, fashion.

Further elucidation of the distinct antigens for these amyloid antibodies has emerged from attempts to identify normal plasma proteins that crossreact with amyloid antibodies. To date, only antibodies of group one, as described above, show unequivocal reaction with specific human plasma proteins (49). The principal crossreactive protein identified with antibodies to amyloid filaments (group 1 antibodies) is a doublet at approximate molecular weights of 88 and 95 kDa. We have partially purified these proteins from human plasma using ammonium sulfate fractionation and Affigel chromatography. Tentative immunochemical identification of the 88-95 kDa bands suggests that they crossreact with antibodies to α-1 antichymotrypsin. This observation was predicated on the fact that the same amyloid antisera that identified these bands were used to screen human liver cDNA libraries identified clones encoding α-1 antichymotrypsin (50). This hepatic-synthesized protein is abundantly present in normal plasma. It is a well-characterized inhibitor of cerine proteases. The identification and cloning of this plasma protein using antibodies to AD amyloid filaments, raises the possibility that α-1 antichymotrypsin may have a role in processing of amyloid proteins, presumably including the beta protein, in AD brain. Several antibodies to α-1 antichymotrypsin immunolabel plaque and vascular amyloid deposits in AD brain sections. However, it is possible that the presence of this protein inhibitor merely represents non-specific sticking of a serum protein to the hydrophobic amyloid filaments. Whether α-1 antichymotrypsin or other protein inhibitors, as well as proteases, play a role in the local processing of the amyloid β-precursors requires further study.

CONCLUSION

Studies to date provide a complex picture of PHF and amyloid filament composition. In the case of the intraneuronal fibers, the microtubule-associated phosphoprotein, tau is an antigenic constituent of the PHF. It is currently not clear to what extent MAP2 and neurofilament proteins also contribute to PHF. Other neuronal proteins, particularly ubiquitin, appear to be present in addition to tau. Thus, the process of neurofibrillary degeneration appears to involve an alteration and insolubilization of certain microtubule-associated proteins together with other constituents that remain to be clarified. Whether this process, which occurs in a number of unrelated neurofibrillary disorders, depends solely on abnormal post-translational processing of cytoskeletal proteins and/or involves changes in transcriptional or translation control of the expression of these proteins remains to be seen.

Current evidence suggests that the brain amyloid filaments present in AD, Down's syndrome and normal brain aging but absent in other NFT-forming disorders (e.g., Guam PD complex, post-encephalitic Parkinson's disease, dementia pugilistica) are distinct from PHF. A major component of amyloid fibrils found both in senile plaque cores and in meningeal and cortical microvascular deposits is a low molecular weight hydrophobic protein (the β-protein) with a previously unknown sequence. Very recent studies from several different laboratories indicate that the gene encoding this amyloid protein is localized to chromosome 21. This observation provides a direct link to the virtually occurrence of senile plaques and microvascular amyloid in late-stage Down's syndrome, although the mechanism of their development with age remains obscure. The finding of amyloid deposits in normal aged humans and in several aged mammals that are immunochemically indistinguishable from AD amyloid suggests a special mechanism for accelerated amyloid deposition in patients with Alzheimer's disease. The precise cellular and tissue origin of the amyloid precursor protien that gives rise to the filaments found in CNS vessels and neuropil is not yet clear. Moreover, the initial identification of amyloid-associated proteins raises the possibility of proteins distinct form the β-protein that could be involved in the cleavage and local processing of the β-protein precursor.

The exciting recent progress in isolating and characterizing AD amyloid components and the genes that encode them raises the likelihood that the amyloid itself may be a seminal factor in cortical degeneration in Alzheimer's disease. Evidence that the amyloid filaments and their proteins can directly induce neuritic reation (and regeneration) and cortical injury (and attempted repairs) is now very important to obtain. Amyloid-bearing neuritic plaques in aged non-human primate cortex may represent a relatively close analog of senile plaques in humans and thus present the prospect of more dynamic studies of the process of senile plaque formation. It remains to be seen to what extent the process of microvascular and cortical amyloid deposition and neuritic plaque formation will explain the etiopathogenesis of Alzheimer's disease. Deciphering the abnormal protein chemistry of AD now enables the generation of specific molecular probes (e.g., complementary DNA's) that can be used to answer fundamental questions about gene expression in this disease and will be valuable in genetic linkage analysis of kindreds with familial AD.

REFERENCES

1. Kidd M: Paired helical filaments in electron microscopy of Alzheimer's disease. Nature 197:192-193, 1963.
2. Terry RD: The fine structure of neurofibrillary tangles in Alzheimer's disease. J Neuropathol Exp Neurol 22:629-642, 1963.
3. Wisniewski HM, Narong HK, Terry RD: Neurofibrillary tangles of paired helical filaments. J Neurol Sci 27:173-181, 1976.
4. Gonatas NK, Anderson W, Evangelista I: The contribution of altered synapses in the senile plaque: an electron microscopic study in Alzheimer's dementia. J Neuropathol Exp Neurol 26:25-39, 1967.
5. Shiboyama H, Kitoh J: Electron microscopic structure of Alzheimer's neurofibrillary changes in a case of atypical senile dementia. Acta Neuropathol (Berl) 41:229-234, 1978.
6. Yagashita S, Itoh T, Nan W, Amano N: Reappraisal of the fine structure of Alzheimer's neurofibrillary tangles. Acta Neuropathol (Berl) 54: 239-246, 1981.
7. Metuzals J, Montpetit V, Clapin DE: Organization of the neurofilamentous network. Cell Tissue Res 214:455-482, 1981.
8. Selkoe DJ. The fibrous cytopathology of degenerating neurons in Alzheimer's disease: an ultrastructural reappraisal. Soc Neurosci Abstr 10:813, 1984.
9. Tomlinson BE, Blessed G, Roth M: Observations on the brain of demented old people. J Neurol Sci 11:205-243, 1970.
10. Wilcock GK, Esiri MM: Plaques, tangles and dementia: A quantitative study. J Neurol Sci 56:343-356, 1982.
11. Ball MJ: Neuronal loss, neurofibrillary tangles and granulovacuolar degeneration in the hippocampus with aging and dementia. A quantitative study. Acta Neuropathol (Berl) 37:111-118, 1977.
12. Perry EK, Tomlinson BE, Blessed G, et al: Correlation of cholinergic abnormalities with senile plaques and mental test scores in senile dementia. Br Med J 2:1457-1459, 1978.
13. Wilcock GK, Esiri MM, Bowen DM, et al: Alzheimer's disease: Correlation of cortical choline acetyltransferase activity with the severity of dementia and histological abnormalities. J Neurol Sci 57:407-417, 1982.
14. Rasool CG, Svendsen CN, Selkoe DJ: Neurofibrillary degeneration of cholinergic and noncholinergic neurons of the basal forebrain in Alzheimer's disease. Ann Neurol 20(4):482-488, 1986.
15. Selkoe DJ: Review: Altered structural proteins in plaques and tangles: What do they tell us about the biology of Alzheimer's disease. Neurobiol Aging 7:425-432, 1986.
16. Kosik KS, Duffy LK, Dowling MM, et al.: Microtubule-associated protein 2: monoclonal antibodies demonstrate the selective incorporation of certain epitopes into Alzheimer neurofibrillary tangles. Proc Natl Acad Sci USA 81:7941-7945, 1984.
17. Selkoe DJ, Ihara Y, Salazar FJ: Alzheimer's disease: insolubility of partially purified helical filaments in sodium dodecyl sulfate and urea. Science 215:1243-1245, 1982.
18. Selkoe DJ, Ihara Y, Abraham C, et al.: Biochemical and immunocytochemical studies of Alzheimer paired helical filaments, in Katzman R (ed): Banbury Report 15: Biological Aspects of Alzheimer's Disease. New York, Cold Spring Harbor Laboratory, 1983, pp. 125-134.

19. Iqbal K, Zaidi T, Thompson CH, et al.: Alzheimer paired helical filaments: bulk isolation, solubility and protein composition. Acta Neuropathol (Berl) 62:167-177, 1984.
20. Selkoe DJ, Abraham C, Ihara Y: Brain transglutamase: In vitro cross-linking of human neurofilament proteins into insoluble polymers. Proc Natl Acad Sci USA 79:6070-6074, 1982.
21. Kirschner DA, Abraham CA, Selkoe DJ: X-ray diffraction from intraneuronal paired helical filaments and extraneuronal amyloid fibers in Alzheimer disease indicates cross-ß conformation. Proc Natl Acad Sci USA 83:503-507, 1986.
22. Ihara Y, Abraham C, Selkoe DJ: Antibodies to paired helical filaments in Alzheimer's disease do not recognize normal brain proteins. Nature 304:727-730, 1983.
23. Perry G, Rizzuto N, Autilio-Gambetti L, et al.: Alzheimer's paired helical filaments contain cytoskeletal components. Proc Natl Acad Sci USA 83:3916-3920, 1985.
24. Brion JP, Couck AM, Passareiro E, et al.: Neurofibrillary tangles of Alzheimer's disease: an immunohistochemical study. J Submicrosc Cytol 17:89-96, 1985.
25. Selkoe DJ, Abraham CR: Isolation of paired helical filaments and amyloid fibers from human brain, in Vallee R (ed), Methods in Enzymology, in press.
26. Wang GP, Grundke-Iqbal I, Kascsak RJ, et al.: Alzheimer neurofibrillary tangles: Monoclonal antibodies to inherent antigen(s). Acta Neuropathol (Berl) 62:268-275, 1984.
27. Selkoe DJ, Abraham C, Rasool CG, et al.: Production and characterization of monoclonal antibodies to Alzheimer paired helical filaments. Ann NY Acad Sci 455:774-778, 1985.
28. Yen S-H, Crowe A, Dickson DW: Monoclonal antibodies to Alzheimer's neurofibrillary tangles. I. Identification of polypeptides. Am J Pathol 120:282-291, 1985.
29. Brion JP, van den Bosch de Aguilar, Flament-Durand J: in, Traber J, Gispens WH (eds.), Berlin, Springer, 1985, pp. 164-174.
30. Nukina N, Ihara Y: One of the antigenic determinants of paired helical filaments. J Biochem 98:1715-1718, 1985.
31. Grundke-Iqbal I, Iqbal K, Quinlan M, et al.: Microtubule-associated protein tau: a component of Alzheimer paired helical filaments. J Biol Chem 261:6084-6089, 1986.
32. Kosik KS, Joachim CL, Selkoe DJ: The microtubule-associated protein, tau, is a major antigenic component of paired helical filaments in Alzheimer's disease. Proc Natl Acad Sci USA 83:4044-4048, 1986.
33. Wood JG, Mirra SS, Pollock NJ, et al.: Neurofibrillary tangles of Alzheimer's disease share antigenic determinants with the axonal microtubule-associated protein tau. Proc Natl Acad Sci USA 83:4040-4043, 1986.
34. Nukina N, Ihara Y: Proteolytic fragments of Alzheimer's paired helical filaments. J Biochem 98:1715-1718, 1985.
35. Drubin DG, Caput D, Kirschner MW: Studies on the expression of the microtubule-associated protein, tau, during mouse brain development, with newly isolated complementary DNA probes. J Cell Biol 98:1090-1097, 1984.
36. Neve RL, Harris P, Kosik KS, et al.: Identification of cDNA clones for the human microtubule-associated protien, tau, and chromosomal localization of the genes for tau and microtubule-associated protein 2. J Molec Brain Res 1:271-280, 1986.
37. Neve R, Selkoe DJ, Kurnit DM, et al.: A cDNA for a human microtubule-associated protein 2 epitope in the Alzheimer neurofibrillary tangle. J Molec Brain Res 1:193-196, 1986.
38. Mori H, Kondu J, Ihara Y: Ubiquitin is a component of paired helical filaments in Alzheimer's disease. Science 235:1641-1644, 1987.

39. Nukina N, Kosik KS, Selkoe DJ: Recognition of Alzheimer paired helical filaments by monoclonal neurofilament antibodies is due to crossreaction with tau protein. Proc Natl Acad Sci USA (in press).
40. Selkoe DJ, Abraham CR, Podlisny MB, Duffy LK: Isolation of low-molecular-weight proteins from amyloid plaque fibers in Alzheimer's disease. J Neurochem 46 (6):1820-1834, 1986.
41. Glenner GG, Wong CW: Alzheimer's disease: Initial report of the purification and characterization of a novel cerebrovascular amyloid protein. Biochem Biophys Res Commun 120:885-890, 1984.
42. Masters CL, Simms G, Weinman NA, Multhaup G, McDonald BL, Beyreuther K: Amyloid plaque core protein in Alzheimer disease and Down's syndrome. Proc Natl Acad Sci USA 82:4245-4249, 1985.
43. Gorevic P, Goni F, Pons-Estel B, Alvarez F, Peress R, Frangione B: Isolation and partial characterization of neurofibrillary tangles and amyloid plaque cores in Alzheimer's disease: Immunohistological studies. J Neuropathol Exp Neurol 45:647-664, 1986.
44. Selkoe DJ, Bell D, Podlisny MB, Cork LC, Price DL: Conservation of brain amyloid proteins in aged mammals and humans with Alzheimer's disease. Science 235:873-877, 1987.
45. Wisniewski HM, Ghetti B, Terry RD: Neuritic (senile) plaques and filamentous changes in aged rhesus monkeys. J Neuropathol Exp Neurol 32:566-584, 1973.
46. Wisniewski HM, Johnson AB, Raine CS, Kay WJ, Terry RD: Senile plaques and cerebral amyloidosis in aged dogs: a histochemical and and ultrastructural study. Lab Invest 23:287-296, 1970.
47. Struble RG, Cork LC, Whitehouse PJ, Price DL: Cholinergic innervation in neuritic plaques. Science 216:413-415, 1982.
48. Struble RG, Price DL, Jr, Cork LC, Price DL: Senile plaques in cortex of aged normal monkeys. Brain Res 316:267-275, 1985.
49. Selkoe DJ, Galibert L, Podlisny M, Duffy LK: Normal human plasma proteins share antigenic determinants with CNS amyloid fibrils in Alzheimer's disease (AD). Soc Neurosci Abstr 12 (1):35, 1986.
50. Abraham C, Selkoe DJ, Potter H: A component of brain amyloid in Alzheimer's disease is the cerine protease inhibitor alpha-1 antichymotrypsin (submitted).

ALTERATIONS OF THE NEURONAL CYTOSKELETON IN ALZHEIMER'S DISEASE AND RELATED CONDITIONS

Khalid Iqbal, Inge Grundke-Iqbal and Henryk M. Wisniewski

New York State Office of Mental Retardation and Developmental Disabilities Institute for Basic Research in Developmental Disabilities, 1050 Forest Hill Road, Staten Island, N.Y.10314

The cytoskeleton of a normal mature neuron is composed of three types of fibrils, the microtubules, the neurofilaments and the microfilaments. One of the cellular and molecular changes with aging, the mechanism of which remains unknown to date, is the formation of argentophilic intracellular neurofibrillary tangles in certain selected neurons of the aged human brain.These extraordinary neurofibrillary changes are seen in great abundance in several adult and late life dementias especially the Alzheimer disease/senile dementia of the Alzheimer type (AD/SDAT) (for review see Iqbal, et al. 1977b; Iqbal and Wisniewski 1983). The Alzheimer neurofibrillary tangles (ANT) are composed of paired helical filaments (PHF). Bundles of these PHF are also found in neurites (Braak, et al. 1986), especially in the dystrophic neurites of the neuritic (senile) plaque, the second leading histopathological lesion of AD/SDAT. Together these two lesions, the ANT and the plaques, both of which contain the PHF, are the histopathological hallmark of AD/SDAT (Kidd, 1964; Terry, et al. 1964); occassionally either tangles of 15 nm straight filaments or these filaments admixed with PHF have been observed in a few AD/SDAT cases. (Shibayama and Kitoch, 1978; Yagishita, et. al. 1981). The number of ANT and plaques correlates positively with the degree of psychometric deficiency in the affected patients (Tomlinson, et. al. 1970), but their origin and role in disease are not understood. The number of tangles and plaques does not appear to be interdependent because in some cases there are numerous tangles and very few plaques and vice versa. In this review the morphology and the biochemistry of the neurofibrillary changes and their relationship to plaque amyloid and normal neurofibrils are discussed.

NORMAL NEUROFIBRILS

The cytoskeleton of a normal neuron comprises of three major neurofibrils, microtubules, neurofilaments and microfilaments. Microtubules of the neuron are apparently identical to those in glial cells and all eukaryotic cells (Olmsted and Borisy, 1973). Each microtubule measures 20-24 nm in diameter, has well defined lumen of about 15 nm and has short side arms. The protein subunit of microtubules is tubulin, a heterodimeric protein. The apparent mol. wts. by sodium dodecyl sulfate-polyacrylamide gel electrophoresis (SDS-PAGE), of human brain tubulin monomers are 56,000 for α and 53,000 for β. These tubulin monomers are acidic and differ slightly in amino acid composition and tryptic and cyanogen bromide peptide maps (Feit, et. al. 1971; Iqbal, et. al. 1977a). About 80-85% of microtubule protein is tubulin, most of the remaining 15-20% protein is of two groups of microtubule associated proteins (MAPS), one group of about 225 kilodalton (kDa) to 350 kDa, called high mol. wt. proteins and the other of about 68 kDa, called "Tau"; by SDS-PAGE tau resolves into several polypeptide bands the most prominent of which correspond to 55 kDa-62 kDa. These various tau molecular species have large homologies as determined by one dimensional peptide maps of their limited proteolysis (Cleveland et al. 1977). In the absence of MAPS microtubules assembled <u>in vitro</u> do not have side arms (Dentler, et. al. 1975; Murphy and Borisy, 1975). Tau is believed to be required for the <u>in vitro</u> assembly of tubulin into microtubules (Weingarten, et. al. 1975). Phosphorylation of tau depresses in vitro microtubule assembly (Lindwall & Cole, 1984). In vitro assembled brain microtubules contain as contaminants small amounts of several brain proteins (Iqbal, et. al. 1977a). The function of microtubules in brain is not yet completely established but they are believed to be involved in the movement of cytoplasmic constituents, especially in axoplasmic flow.

The neurofilaments are the intermediate filaments of the neuron. They are linear, 9-10 nm in diameter and have side arms. They are found sparsely in the cell body, moderately in the dendrites and most abundantly in the axon. Like microtubules, neurofilaments can be made to undergo <u>in vitro</u> disassembly-assembly cycles (Iqbal, et. al. 1981; 1986a;1987d). However, the conditions for disassembly and assembly of neurofilaments are different from those of microtubles. Neurofilaments are biochemically different from intermediate filaments of other cell types and are made up of a triplet of about 70 kDa (neurofilament light, NF-L), 160 kDa (NF-medium, NF-M) and 200 kDa (NF-heavy, NF-H) polypeptides that are apparently unique to nerve cells (Hoffman and Lasek, 1975; Schlaepfer, 1977; Soifer, et al. 1981). Neurofilaments isolated from CNS contain varying amounts of a 50 kDa polypeptide which is believed to be mostly the glial fibrillary acidic protein, the major protein of astroglial filaments (Eng, et.al. 1971). The function of neurofilaments is not yet understood. They may act as a part of the force-generating mechanism in axoplasmic transport, and as a cytoskeletal element in the preservation of cell asymmetry.

Brain microfilaments, like those in muscle are 5 nm in diameter and are made up of actin, the 45 kDa polypeptide (Berl, et. al. 1973).

NEUROFIBRILLARY CHANGES IN AD/SDAT

Both ultrastructure (Table I) and biochemistry of PHF (Table 4) have been an area of controversey. Although AD/SDAT and the ANT were discovered by Alzheimer as early as in 1907, most of the studies on ultrastructure and biochemistry of PHF were reported in the last few years.

<u>Morphology of ANT</u>:

In the light microscope, ANT appears as an intraneuronal mass which is stained intensely with silver impregnation techniques. It produces green birefringence in polarized light after staining with congo red. This optical property is not observed in normal neurons and is believed to be due to β-pleated sheet nature of the proteins making the ANT (Glenner, et al. 1974). X-ray diffraction patterns of isolated ANT are indeed suggestive of β-pleated proteins (Kirschner, et.al. 1984).

TABLE 1: HISTORY AND CONTROVERSIES ON ULTRASTRUCTURE OF PHF

Authors	Ultrastructure of PHF
Kidd, 1963	PHF; made from two 10-13 nm filaments helically wound around each other
Terry, 1963	Twisted tubules, tubular structure with intermittent constrictions
Hirano, et al. 1968	Twisted ribbons
Wisniewski, et al. 1976	PHF spiral substructure
Wisniewski, et al. 1984	PHF, made up of 4-8 protofilaments each 3-5 nm in diameter
Brion, et al. 1984	PHF, right handed spiral substructure formed by three 4-6 nm strands; each strand contain paired 2-3 nm protofilaments
Wisniewski and Wen, 1985	PHF, 8 protofilaments each consisting of globules (32°A $\pm$ 4) connected by longitudinal bars (47°A $\pm$ 6); different from neurofilaments
Wischik, et.al. 1985	Left handed twisted ribbons
Wisniewski, et al. 1986	PHF, two types: right handed (RH) and left handed (LH) helices; RH are larger than the LH both in diameter and in periodicity

Ultrastructurally PHF are unlike any of the normal neurofibrils (Table I). Each PHF is a pair of filaments, 10-13 nm in diameter wound helically around each other at regular intervals of 80 nm (Kidd, 1963; Wisniewski, et. al. 1976; Wisniewski, et al. 1984). Each PHF is made up of eight protofilaments; in longitudinal section only four protofilaments are seen, the other four are hidden behind. The substructure of PHF is different from normal neurofilaments in that the globules making the PHF protofilaments are larger (32° A $\pm$ 4 vs 20° A $\pm$ 3) and the longitudinal bars are longer (47° A $\pm$ 6 vs 27° A $\pm$ 2) than those in neurofilaments (Wen & Wisniewski, 1984; Wisniewski and Wen 1985). According to Wischik, et al. (1985), PHF are twisted ribbons wound in a left handed manner.

In addition to ANT, PHF are found as bundles in the dystrophic neurites of the senile plaques and in neurites in the neuropil of the telencephalic cortex. At this third location, PHF occur in small and inconspicuous profiles scattered throughout both allocortical and isocortical areas (Braak, et. al. 1986). The patterns of distribution and packing density of neuropil threads i.e. neuropil containing accumulations of PHF varies between different cortical areas and layers. In neurons undergoing neurofibrillary changes, PHF appear to gradually become more densely packed and take over greater proportions of cell space displacing cytoplasmic organelles. It remains unclear whether accumulations of PHF lead to cell death. It is also unknown whether the affected cells can recover. Maintenance of synaptic contact has been observed in situations where pre and post synaptic processes are filled with PHF suggesting that a certain degree of function might persist in in these affected synapses.

Topography of ANT

ANT are found mostly in cerebral cortex, especially in the hippocampal pyramidal neurons of Sommers sector and in small pyramidal neurons in the outer laminae of fronto-temporal cortex. They have not been observed in cerebellar cortex, spinal cord, peripheral nervous system or extraneuronal tissues.

NEUROFIBRILLARY CHANGES IN CONDITIONS OTHER THAN AD/SDAT

In addition to AD and SDAT which are believed to be the same disease with a different age of onset, PHF are also found in great abundance in Guam Parkinsonism dementia complex, dementia pugilistica, postencephalitic Parkinsonism and adults with Down syndrome (for review see Iqbal, et.al. 1977b; Wisniewski et. al. 1979; Iqbal and Wisniewski, 1983). The ANT has also been reported in small numbers in several cases of subacute sclerosing panencephalitis (SSPE) and in rare cases of Hallerworden Spatz disease and juvenile neurovisceral lipid storage disease (see Table 2). The PHF in these different human conditions are morphologically identical to that in AD/SDAT and antisera to PHF isolated from AD/SDAT brain label PHF in aged brain, Down syndrome, Guam-Parkinsonism dementia complex and AD/SDAT cases (Grundke-Iqbal, et. al. unpublished observations). Thus the accumulation of PHF is associated with normal aging, viral infection, chromosomal disorders and metabolic abnormalities. However, PHF of the Alzheimer type have never been observed in any aged animal species or have they been produced experimentally in animals.

The neurofibrillary changes in human disorders are not always of the Alzheimer type i.e. made up of PHF. For instance in progressive supranuclear palsy (PSP) neurons of some of the same areas which contain tangles of PHF in Alzheimer brain have neurofibrillary tangles of 15 nm straight filaments (Tellez-Nagel and Wisniewski, 1973). These tangles of 15 nm filaments in PSP are sometimes admixed with PHF (Ghatak, et.al. 1980). The 15 nm filaments crossreact immunochemically with PHF since antisera to isolated PHF and as well as PHF reactive antisera to tau (see below) label PSP tangles at both light and electron microscopic level in tissue sections (Grundke-Iqbal, et. al. 1982; Pollock, et al. 1986; Gorevic, et.al. 1986; Bancher, et al. 1987).

In sporadic motor neuron disease, vincristine neuropathy and infantile neuroaxonal dystrophy in humans the neurofibrillary changes are of 10 nm intermediate filament type. The 10 nm intermediate filament type neurofibrillary changes have been experimentally induced with aluminum, mitotic spindle inhibitors like colchicine, vinblastine and podophyllotoxin, various nitrates and acrylamide in animals. The aluminum-induced filamentous accumulation is apparently specific to the nervous system, while in case of the mitotic spindle inhibitors similar changes occur in a wide range of cell types. In all of these intoxications, the filaments formed are morphologically identical to normal neurofilaments. These filaments induced with various agents have not yet been isolated and characterized. The neurofilamentous tangles induced with aluminum, colchicine, vinblastine and vincristine have been shown to immunostain with an antiserum to the 70 kDa neurofilament polypeptide (NF-L) but not with an antiserum to the 160 kDa neuro-filament polypeptide (NF-M) (Dahl and Bignami, 1978; Dahl, et. al. 1980). Unlike PHF, both normal neurofilaments and the aluminum-induced

TABLE 2: CONDITIONS AND TYPE OF NEUROFIBRILLARY CHANGES

Condition	PHF	10 nm Filament
Human:		
Aged persons	+[a]	
Alzheimer presenile & senile dementia	+[b]	
Guam Parkinsonism dementia	+[b]	
Dementia pugilistica	+[b]	
Down Syndrome	+[b]	
Post encephalitic parkinsonism	+[b]	
Subacute sclerosing pan encephalitis (SSPE)	+[c]	
Hallervorden-Spatz disease	+[c]	
Progressive supranuclear palsy (PSP)	+[d], 15 nm	
Sporadic motor neuron disease		+[e]
Vincristine neuropathy		+[e]
Infantile neuroaxonal dystrophy		+[e]
Animal:		
Aged Rhesus monkey	**	
Chronic alcohol treated rat	**	
Whip spider	**	
Aluminum encephalomyelopathy		+[e]
Spindle inhibitor encephalopathy		+[e]
Lathyrogenic encephalopathy (IDPN)		+[e]
Vitamin E deficiency		+[e]
Copper deficiency		+[e]
Retrograde and Wallerian degeneration		+[e]

[a]Small numbers of tangles
[b]Numerous tangles
[c]Small number of tangles and in a few cases only
[d]Tangles of 15 nm straight filaments some of which are admixed with various amounts of PHF
[e]Tangles of 10 nm straight filaments which are morphologically identical to normal neurofilaments.
**PHF found in these conditions are sparse and have different dimensions from that of Alzheimer presenile and senile dementia

filaments do not produce in polarized light the characteristic green birefringence after staining with congo red.

MORPHOLOGICAL AND BIOPHYSICAL HETEROGENEITY OF ANT AND PHF

ANT are heterogenous both in their morphology and solubility. Most of the ANT are made up of PHF. Only a few ANT are a mixture of PHF and straight 10-15 nm filaments.

PHF are stable in both fresh and frozen autopsy tissue and are resistant to solubilization in aqueous buffer in the absence of detergents or denaturants. Two general populations of ANT, I and II have been identified based on the solubility and insolubility,

TABLE 3: METHODS FOR BULK ISOLATION OF PHF

Authors	Method
Iqbal, et al. 1974	Nondenaturing conditions; isolation of neuronal perikarya followed by conventional subcellular fractionation.
Grundke-Iqbal, et al. 1981* Iqbal, et al. 1984	Isolation of neuronal perikarya followed by treatment with 2% SDS at room temperature for 3-5 min. and sucrose density gradient centrifugations. Yield: 10-100 ug protein/g tissue.
Selkoe, et.al. 1982	Differential centrifugation of tissue homogenate, followed by extraction in 2% Triton X-100; yield: 1-2 mg protein/g tissue.
Ihara, et.al. 1983	Boiling of tissue homogenate in 2% SDS and 0.1 M β-mercaptoethanol buffered with 0.05 M Tris, pH 7.6; Yield 50-80 ug/g tissue.
Masters, et.al. 1985b	Extraction of tissue homogenate with salts and Triton X-100, followed by digestion with pepsin, sucrose discontinuous density gradient centrifugations and extraction with 2% SDS.
Rubenstein, et.al. 1986	Differential and rate zonal centrifugation of tissue hommogenate digested with proteinase-K and micrococcal nuclease and extracted by sonication with sarcosyl and sulfobetain 3-14; yield: 0.2 ug/g tissue.

*Isolation of PHF employing treatment with SDS, polypeptide composition of isolated PHF and aggregation of PHF polypeptides at the top of the gel were presented for the first time at the Annual Meeting of the American Assoc. Neuropathologists in 1981.

respectively in 2% SDS at room temperature for 3-5 minutes (Iqbal, et.al. 1984). The ANT II are, however, solubilized on repeated extractions in SDS and β-mercaptoethanol at 90-100°C or more effectively by ultrasonication followed by heating in 1% each of SDS and β-mercaptoethanol) (Iqbal, et.al. 1984; Iqbal, et.al. 1986b).

Highly purified PHF are isolated from autopsied tissue by a combination of sucrose density gradient centrifugation and SDS treatment of neuronal cell bodies enriched preparations (Grundke-Iqbal, et al. 1981;Iqbal, et. al. 1984). PHF isolated by heating the whole chopped tissue with SDS and β-mercaptoethanol according to Ihara, et. al. (1983) are less purified and are made insoluble most likely with this harsh detergent treatment (Iqbal, et. al. 1986b; Rubenstein, et.al. 1986).

While native PHF are resistant to proteolysis, PHF isolated by SDS-treatment are readily digested with proteases. Solubility of PHF in detergents and denaturants and their proteolysis demonstrate that, contrary to the report of Selkoe, et. al. (1982), PHF are not completely insoluble polymers of polypeptides crosslinked with γ-glutamyl-ε-lysine.

In agreement with the biochemical studies, ultrastructural studies of PHF have subsequently revealed that, indeed, there are two general populations of PHF, i.e., PHF with right-handed helices and PHF with left-handed helices (Wisniewski, et. al. 1986). The right-handed PHF are larger than the left-handed PHF both in diameter and in periodicity of the helices. Furthermore, it has also been observed that the right-handed PHF are more readily soluble than the left-handed PHF by detergent treatments.

POLYPEPTIDE COMPOSITION OF PHF

The very first report on the isolation of PHF, which formed the basis of all subsequent methods for isolation of PHF, and identification of a 50 kDa PHF polypeptide (see Tables 3 and 4), was made in 1974 (Iqbal, et.al. 1974). Since no ultrasonication or repeated heat extractions in SDS and β-mercaptoethanol were employed for the preparation of PHF enriched samples for gel electrophoresis at that time, the 50 kDa PHF polypeptide might represent mostly the ANT I/right-handed PHF. Antibodies to this 50 kDa PHF polypeptide were subsequently shown to label ANT and plaque neurites which also contain PHF, and form in Ouchterlony immunodiffusion test a line of identity with PHF and *in vitro* assembled normal brain microtubules (Grundke-Iqbal, et.al. 1979a). Furthermore, antisera raised against *in vitro* assembled microtubules from normal brain were shown to label ANT and plaque neurites (Grundke-Iqbal, et.al. 1979b). In fact this was the first demonstration of a crossreactivity of a normal cytoskeletal protein/s with PHF. Subsequently a number of reports confirmed the crossreactivity of PHF with one or another normal cytoskeletal protein (Ishii, et.al. 1979; Yen, et. al. 1981, 1983; Anderton, et.al. 1982; Dahl, et.al. 1982; Gambetti, et.al.1980; 1983; Kosik, et.al. 1984; Perry, et.al. 1985; Sternberger, et.al. 1985). However, as it has been pointed out previously (Grundke-Iqbal, 1986a) the immunocytochemical crossreactivity between two antigens does not allow to draw any precursor product relationships. The size of an antigenic site detected by an antibody is relatively small and identical or closely related antigenic sites comprised of a few amino acids have been found on molecules which are otherwise unrelated, e.g., between the transforming proteins of Rous sarcoma virus and tubulin, myosin, and vimentin (Nigg, et.al. 1982), between thymus and neuronal cell surface proteins (Golub and Day, 1975) and between rabbit IgG and ANT (Grundke-Iqbal, et.al. 1985c). Furthermore, it is not possible to determine with immunocytochemical techniques alone whether crossreactivities may be due to similarities in primary amino acid sequence or due to antigenic sites created from folding of different regions of a polypeptide chain (Berzofsky, et.al. 1982).

To date only microtubule associated protein tau has been shown to be a component of PHF (Grundke-Iqbal, et.al. 1986a, 1986b). The polypeptide composition of PHF is heterogeneous. The major polypeptides of isolated PHF have been identified as 45 kDa-62 kDa by SDS-PAGE (Iqbal, et.al. 1984) and Western blot analysis of isolated PHF developed with monoclonal and polyclonal antibodies to PHF (Grundke-Iqbal, et.al. 1984, 1985b). These 45 kDa-62 kDa PHF polypeptides are the same as tau by polypeptide compositional analysis and their Western blots developed with monoclonal and polyclonal antibodies to tau and to PHF

TABLE 4: HISTORY AND CONTROVERSIES ON POLYPEPTIDE COMPOSITION OF PHF

Authors	Major findings
Iqbal, et al. 1974	A 50 kDa polypeptide identified in a PHF-enriched fraction.
Iqbal, et.al. 1978	The tyrosine containing tryptic and chymotryptic peptide maps of the 50 kDa PHF polypeptide have many similarities to the peptide maps of the 50 kDa polypeptide from brain filament enriched fractions and 53 kDa and 56 kDa polypeptides (from SDS-PAGE, most likely contaminated with tau) corresponding to α and βtubulins from microtubule enriched preparations.
Grundke-Iqbal,et.al. 1979a	Antisera to the 50 kDa PHF polypeptide labelled ANT and plaque neurites in tissue sections, and formed, in Ouchterlony test, a line of identity with the PHF polypeptide and microtubles.
Grundke-Iqbal,et.al. 1979b	Antisera to normal brain microtubules labelled ANT and plaque neurites in tissue sections.
Selkoe, 1980	A 20 kDa (and not 50 kDa) polypeptide identified in a PHF-enriched fraction.
Selkoe, et.al. 1981	The 20 kDa polypeptide in AD/SDAT neuronal (PHF) enriched fraction is myelin basic protein.
Selkoe, et.al. 1982	PHF insoluble in SDS and other denaturants; their polypeptide composition cannot be studied; hypothesized that PHF are insoluble polymers made up of polypeptides crosslinked by γ-glutamyl-ε -lysine.
Ihara, et.al. 1983	Antisera to PHF do not recognize any normal cytoskeletal polypeptide.
Iqbal, et.al. 1984 (Grundke-Iqbal, et.al. 1981*)	Two populations of ANT, I and II; I: readily soluble in 2% SDS; II: require for solubility ultrasonication prior to heat extraction in 1% SDS; polypeptide composition heterogeneous, major PHF polypeptides on SDS-PAGE: 45 kDa-62 kDa.
Grundke-Iqbal,et.al. 1981*,1984	Antisera to isolated PHF label the 45 kDa -62 kDa PHF polypeptides.
Wang, et.al. 1984	Production of monoclonal antibodies to PHF.

TABLE 4 Cont'd

Authors	Major findings
Grundke-Iqbal, et.al. 1985b	The 45 kDa-62 kDa PHF polypeptides are labelled with monoclonal antibodies to PHF.
Masters, et.al. 1985b	PHF are made from the same protein (β-protein) as the cerebravascular and plaque core amyloids except that it has ragged N-termini; βprotein in PHF have 7 ragged sequences.
Grundke-Iqbal, et.al. 1985a	ANT-reactive antisera to microtubules label the same 45 kDa-62 kDa PHF polypeptides as stained by anti PHF sera; The ANT-reactive antibodies in the anti microtubule sera are absorbed with PHF polypeptides and microtubule polypeptides (tau region) other than tubulin and high mol. wt. MAPS.
Brion, et.al. 1985; Flament-Durand and Brion, 1986	Antisera to tau label ANT in tissue sections and antisera to PHF label tau.
Grundke-Iqbal,et.al. 1986a	Tau is a major component of PHF: antisera to microtubules and affinity purified polyclonal antibodies to various molecular species of tau label ANT and plaque neurites and the same 45 kDa-62 kDa PHF polypeptides as stained by the polyclonal and monoclonal antibodies to PHF; PHF-staining antibodies in anti microtubule sera are selectively absorbed with tau.
Kosik, et.al. 1986	Antisera to PHF label tau; affinity purified polyclonal antibodies to tau label ANT and plaque neurites in tissue sections.
Grundke-Iqbal, et.al. 1986b,1986c	Tau in PHF is abnormally phosphorylated: PHF polypeptides (45 kDa-62 kDa) and tau from Alzheimer brain cytosol are labelled by a specific monoclonal antibody tau-1 to tau only when dephosphorylated prior to immunostaining; the number of ANT and plaques (neurites only) labelled with tau-1 is dramatically increased when tissue sections are dephosphorylated prior to immunostaining; antisera to PHF label tau.
Wood, et al. 1986	The number of ANT and plaques (neurites only) labelled with the monoclonal antibody to tau-1 is very much increased when tissue sections are dephosphorylated prior to immunostaining.

TABLE 4 Cont'd

Authors	Major findings
Ihara, et.al. 1986	Antisera to PHF label phosphorylated forms of tau.
Delcourte and Defossez, 1986	Antisera to PHF label tau and antisera to tau label ANT.
de Garcini, et.al. 1986	Dialysis of tau from buffered 6 M urea to buffer alone produce self assembly of tau into filaments resembling PHF.
Gorevic, et.al. 1986	PHF are insoluble in SDS and other denaturants; 4 kDa-6 kDa, 15 kDa-17 kDa and 24 kDa polypeptides are extracted from PHF enriched fractions with formic acid/chloroform extraction.

*Sparing solubility of PHF isolated by SDS treatment, their polypeptide composition and immunolabeling of the PHF polypeptides were presented for the first time at the Annual Meeting of the American Association of Neuropathologists in 1981.

(Grundke-Iqbal, et.al. 1985b, 1986a; in preparation). Furthermore, both by immunocytochemistry and by Western blot analysis it has been shown that tau in PHF is abnormally phosphorylated (Grundke-Iqbal, et.al. 1986b).

The amino acid composition of tau and PHF is similar but distinct (Iqbal, et. al. 1986c). These differences might be due to the different degrees of purity of the two sources of the protein, post-translational modifications, differences in the amino acid sequence of the two proteins or due to the presence of additional polypeptides in PHF. It is not currently known what are, in addition to tau, the other components of PHF. Quantitation of tau in PHF by a competitive ELISA (enzyme-linked immunosorbent assay) has revealed that most of the PHF protein has tau immunoreactivity (Grundke-Iqbal, et. al. 1987). In our experience (Grundke-Iqbal, et al., in preparation) antibodies to tau label in tissue sections intraneuronal tangles but not what are believed to be ghost tangles or tombstones left after degeneration of the affected neurons. Furthermore, while both monoclonal and polyclonal antibodies to PHF label the same PHF polypeptides, only the polyclonal but not the monoclonal antibodies to PHF label normal tau on Western blots. These studies suggest that tau in PHF is altered creating new epitopes and that this alteration might be responsible for the inaccessibility of some of normal tau antigenic sites with maturation of PHF.

BRAIN MICROTUBULE ASSEMBLY DEFECT IN AD/SDAT

One of the vital functions of neuron is the transport of materials between the cell body and the nerve endings. Microtubule assembly which is necessary for this intracellular transport called axoplasmic transport is defective in AD/SDAT (Iqbal, et.al. 1986a, 1987a). Interestingly DEAE-dextran, a polycation which mimics tau in stimulating microtubule assembly induces the assembly of microtubules <u>in vitro</u> in Alzheimer brain cytosol (Iqbal, et.al. 1986a). Since tau in AD/SDAT brain cytosol is abnormally phosphorylated (Grundke-Iqbal, et.al. 1986b;

Iqbal. et.al. 1986a) and phosphorylation of tau depresses microtubule assembly (Lindwall and Cole, 1984), it would appear that this alteration of tau in AD/SDAT brain might be responsible for the microtubule assembly defect.

The concentrations of tubulin might drop with age (Yan, et.al. 1985). As is the case in vitro, a critical concentration of brain tubulin is probably required for in vivo microtubule assembly. Any change in tubulin or in microtubule associated proteins that would decrease the efficiency of microtubule assembly, would be thus critical in aged brains. Abnormally phosphorylated tau might thus mean that this threshold is reached in AD/SDAT, the result being reduced microtubule assembly and consequently impaired axoplasmic flow. Since tau in PHF is abnormally phosphorylated, it seems that the altered tau might not be catabolized, at least actively, thereby accumulating as PHF in the affected neurons. A disturbance in axoplasmic flow, both anterograde and retrograde, should lead to accumulations of components of the axoplasmic flow in both the perikaryon and the nerve terminals. PHF accumulate at both of these locations i.e. ANT and plaque neurites. The accumulation of the affected proteins will depend on the rate of their transport, synthesis and the rate of their degradation by the cell. Thus several neuronal components that are normally transported between the cell body and the terminals and not rapidly degraded will be expected to accumulate in the affected neurons. However, only one or a few of these polypeptides might be capable of polymerizing into PHF. Sensitive immunocytochemical techniques have shown that antibodies to several polypeptides (Ishii, et.al. 1979; Anderton, et.al. 1982; Dahl, et.al. 1982; Gambetti, et.al.1980, 1983; Yen, et al. 1983; Grundke-Iqbal, et.al. 1985c; Kosik, et.al. 1984; Perry, et.al. 1985; Roberts, et.al. 1985; Sternberger, et.al. 1985), including tau (Brion, et.al. 1985; Kosik, et.al. 1986; Wood, et.al. 1986; Nukina, et.al. 1986; Delcourte, et al. 1986; Yen, et.al. 1986), attach to ANT and plaque neurites, but to date only tau has been shown, by polypeptide composition analysis, to be a component of PHF (Grundke-Iqbal, et al. 1986a, 1986b).

RELATIONSHIP BETWEEN NEUROFILAMENTS AND PHF

Neurofilaments have been suggested by several investigators to be the source of origin of PHF based only on the immunocytochemical reactivity of PHF with antibodies to neurofilaments (Ishii,et al. 1979; Anderton, et.al. 1982; Dahl, et.al. 1982; Gambetti, et.al.,1980, 1983; Perry, et.al. 1985; Sternberger, et.al. 1985). The only other data in support of the neurofilamentous origin of PHF are the Bodian silver staining of neurofilament polypeptides (Gambetti, et.al. 1981). Bodian silver impregnation on tissue sections is known to stain PHF, neurofilaments and nuclei. Recently we have observed that a silver impregnation technique (Gallya stain) which specifically stains PHF and not neurofilaments, readily stains PHF and tau polypeptides but not neurofilament triplet (Iqbal, et.al. 1987b). All attempts in our lab to demonstrate neurofilament polypeptides in isolated PHF have been, however, unsuccessful, to date (Grundke-Iqbal, et.al. 1984;1985a; 1985b). Furthermore, studies on the substructure of PHF and neurofilaments have also shown that PHF are not the product of helical winding of neurofilaments since both the size of the globular subunit of the protofilaments and distance between the adjacent globules in PHF is larger than that in neurofilaments (Wen and Wisniewski, 1984; Wisniewski and Wen, 1985). Therefore, presently the case for the neurofilament origin of PHF is weak.

PLAQUE AMYLOID

Relationship between PHF (ANT) and plaque amyloid is summarized in Table 5. Both PHF and amyloid are congophilic and because of this common staining property a close relationship between the two lesions has been suggested by some investigators. It has been shown that this property is most likely due to β-pleated sheet structure (Glenner, et. al. 1974) and this conformation can be induced in many unrelated polypeptides which will then in turn become congophilic. PHF and amyloid fibrils are different ultrastructurally and in several staining properties other than with congo red and thioflavin S.

TABLE 5: COMPARISONS BETWEEN PHF AND PLAQUE AMYLOID

Characteristic	PHF	Amyloid
Fibril diameter (nm)	10-22	7
Congophilia	+	+
Argentophilia	+	-
Periodic acid Schiff (PAS)	-	+
Solubility in SDS	+	-
Mol. wt. (SDS-PAGE)	45k-62k	4k
Reaction with anti-PHF sera	+	rare
Reaction with anti-amyloid*	_	+

*Antisera to two synthetic peptides consisting of first 10 and 24 amino residues of amyloid β-protein of Glenner and Wong (1984).

Glenner and Wong (1984 a,b) have isolated and sequenced a 4.2 kDa protein called β-protein from cerebrovascular amyloid from AD/SDAT and Down syndrome brains. The amino acid sequence of the β-protein reported by these investigators is unique and homologous (except one amino acid substitution difference) between AD/SDAT and Down syndrome brains. Oligonucleotide probes have been made from published sequence of the β-protein and successfully employed to isolate the amyloid gene and show its localization on chromosome 21 (Robakis, et.al. 1987; Goldgaber, et.al. 1987, Tanzi, et.al. 1987; Kang, et.al. 1987). Masters, et.al. (1985a,b) have reported that the senile plaque core amyloid and PHF are also made up of the β-protein but have ragged amino-terminal sequences. These authors have claimed 4 and 7 sequences for the plaque amyloid and PHF respectively. However, the difficulty with these data are that the authors have not reported the yield of the protein sequenced from their plaque amyloid and PHF preparations. It is not possible to judge whether they sequenced any major PHF and plaque amyloid proteins or the cerebrovascular amyloid present as a contaminant of their preparations. Masters, et.al. (1985b) have also reported the immunoreactivity with ANT of an antiserum to a synthetic peptide corresponding to the residues 1-11 of the -protein, whereas others (Wong, et.al. 1985; Allsop, et.al. 1986; Selkoe, et.al. 1986; Grundke-Iqbal, et al. unpublished observations) using both monoclonal and polyclonal antibodies to the synthetic β-peptides stained only cerebrovascular and plaque core amyloid but not tangles. Anti PHF sera which label PHF do not by and large label plaque amyloid. PHF staining antibodies in the anti PHF sera are not absorbed with the synthetic amyloid polypeptide (24 amino acid residue) (Grundke-Iqbal, et. al., unpublished data). According to Shirahama, et. al. (1982) both ANT and plaque amyloid are immunostained with a commercially available anti prealbumin serum. These findings have not been reproduced (Alafuzoff, et. al. 1985; Stam and Eikelenboom, 1985). Furthermore, there is no cross- reactivity between PHF and scrapie associated fibrils (SAF) or prion rods (Merz, et.al. 1986); the SAF crossreact with scrapie but not with AD/SDAT plaque amyloid.

Although AD/SDAT is characterized histopathologically by the presence of numerous tangles and plaques in the brain, the two lesions might not be necessarily interdependent. In Guam Parkinsonism dementia, dementia pugilistica and postencephalitic Parkinson disease, numerous ANT are seen with similar topography as in AD/SDAT and as yet by and large no plaques are seen in these disorders. Conversely in slow unconventional viral conditions both in human, i.e., Creutzfeldt-Jakob disease, Kuru and Gerstmann-Straussler syndrome and in animals, i.e., various strains of scrapie infected mice both amyloid and neuritic plaques are seen but without neurofibrillary tsangles. Furthermore even in AD/SDAT in some cases there are numerous ANT and very few plaques and vice versa. Both tangles and plaques are of importance on their own and the understanding of their pathogenetic mechanisms which might or might not be interdependent will enhance our understanding of cerebral aging and AD/SDAT.

SCRAPIE ASSOCIATED FIBRILS (SAF) OR PRION RODS

Scrapie associated fibrils (SAF) (Merz, et. al. 1981) or Prion rods (Pruisner, et. al.1982) are seen in vitro in scrapie-infected brain homogenates fractionated by treatment with proteases and detergents; the SAF are not seen in situ. These fibrils have been recently isolated and scrapie infectivity has been reported to be associated with them (Diringer, et. al. 1983, Prusiner, et. al. 1982; 1983). Since like PHF and plaque amyloid, SAF are congophilic, Pruisner et al. (1983) have proposed that SAF is the infectious agent and that AD/SDAT might be caused by a transmissible agent similar to scrapie. However, it should be pointed out here that congophilia is an indication of β-pleated sheet conformation and can be induced in unrelated polypeptides (Glenner, et. al. 1974). Furthermore, SAF are most likely amyloid fibrils which are formed in vitro during tissue fractionation and are not the scrapie agent; Bendheim, et.al. (1984) and Wisniewski, et al. (1985) have shown the reactivity of anti SAF serum with plaque amyloid in scrapie infected animals. SAF and PHF are distinct from each other in morphology, polypeptide composition and immunochemical crossreactivity (see Table 6) Furthermore, while SAF are observed in Creutzfeld-Jacob disease (Merz, et al. 1983) they are not seen in AD/SDAT.

TABLE 6: COMPARISON BETWEEN PHF AND SCRAPIE ASSOC. FIBRILS (SAF) or PRION RODS (PR)

Characteristic	PHF	SAF (Pr)
Fibril diameter (nm)	10-22nm	5-14
Congophilia	+	+
Argentophilia	+	-
Periodic acid schiff (PAS)	-	prob.+
Solubility in SDS	+	+
Mol. wt. (SDS-PAGE)	45k-62k	25k-30k
Reaction with anti-PHF	+	-
Reaction with anti-SAF	-	+

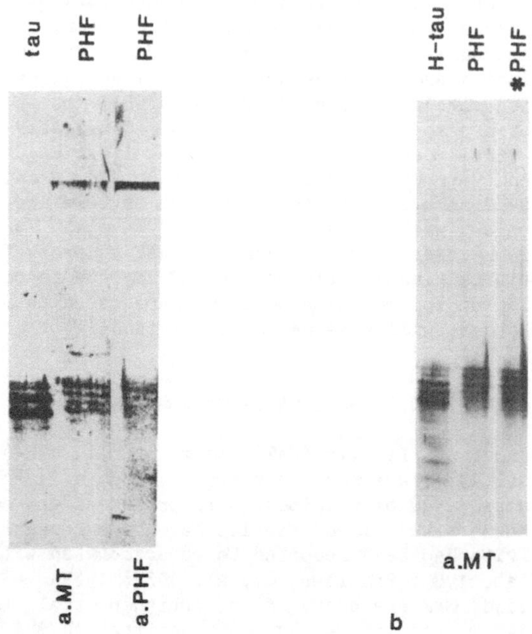

Fig. 1. Comparison of human and calf-brain tau and PHF polypeptides. (a) Calf brain tau labelled with anti-MT (PHF) serum (a. MT) (1:3000) and PHF polypeptides labelled with the same antiserum (a. MT) and with monoclonal antibody 5-25 to PHF (a. PHF) (1:50,000). (b) Human brain tau (H-tau) and PHF polypeptides dephosphorylated (*) or untreated, labelled with anti-MT (PHF) serum (a. MT).

Calf brain tau was prepared by heat treatment at pH 2.7 from thrice cycled microtubules. Human tau was similarly prepared from twice cycled microtubules from the brain of a normal 67 year-old patient. Electrotransfer from SDS gel, 5-15% and 5-10% acrylamide (8 x 6 cm) as shown in a and b respectively. For SDS-polyacrylamide gel electrophoresis, the isolated PHF (1 mg/ml) were solubilized by first sonicating in 0.32 M sucrose for 30 minutes at setting 1 and 10% pulse with 4 seconds on/off cycles using a Branson Model 200 sonifier equipped with a tapered microprobe and then by heating in a boiling water bath for 3 minutes with SDS and β-mercaptoethanol at both 1% final concentration (Iqbal, et.al. 1984). *The paperblot was treated with alkaline phosphatase before application of anti-MT (PHF) serum. Although not shown in this figure, both the 200 and the 150-KDa neurofilament polypeptides employed as control were dephosphorylated under these conditions. Reproduced with permission from Grundke-Iqbal, et.al., J. Biol. Chem. 261:6084-6089, 1986.

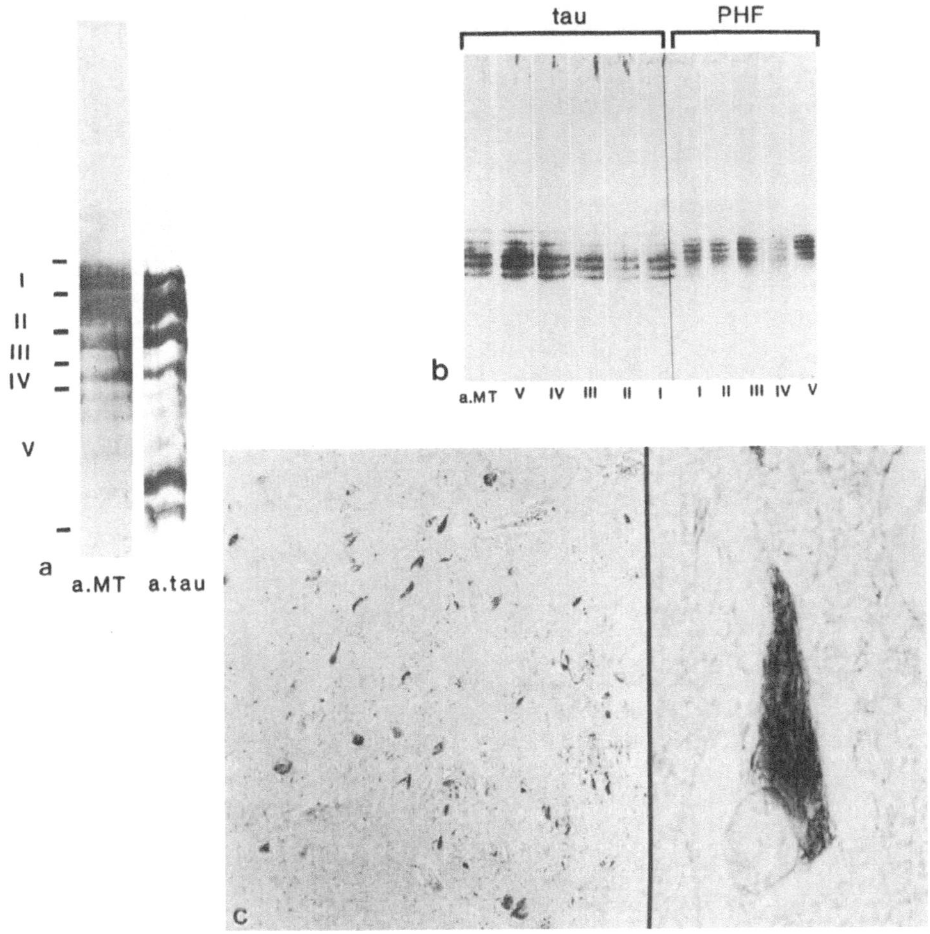

Fig. 2. Labelling of tau and PHF polypeptides on Western blots and of tangles and plaque neurites on tissue sections with antibodies purified by immuno affinity from 5 different molecular species of tau. For tau (prepared by heat treatment at pH 2.7) was electrophoresed on SDS-polyacrylamide gels (12.5% acrylamide, 16 x 11 cm) and transferred to nitrocellulose paper. Strips were cut from the sides and developed with anti-MT (PHF) serum (a. MT) and monoclonal antibody to tau (a. tau). The remaining blot was used for affinity isolation of anti-tau antibodies from the anti-MT (PHF) serum. Roman numerals (I-V) indicate the areas of tau species from which the antibodies were purified. In immunoblots of tau and PHF polypeptides with affinity purified antibodies from tau polypeptides of areas I-V are shown. Electrotransfer from SDS gel, 7-10% acrylamide (8 x 6 cm). Differences in the staining intensities of the different antibodies are due to small individual variations in the amounts of the samples applied to the gel shows immunocytochemical staining of tangles and neurites of plaques in paraffin sections of Alzheimer hippocampus with antibodies eluted from tau area I (left panel); the background staining might correspond to the normal distribution of tau. The right panel shows at high magnification a neuron with the fibrils of its tangle darkly stained by the antibody. Original magnifications: left panel, x 130; right panel, x 1,500. Identical staining was obtained with antibodies eluted from the other 4 tau areas; plaque amyloid was not stained with any of these 5 antibodies. Reproduced with permission from Grundke-Iqbal, et.al., J. Biol. Chem. 261:6084-89, 1986.

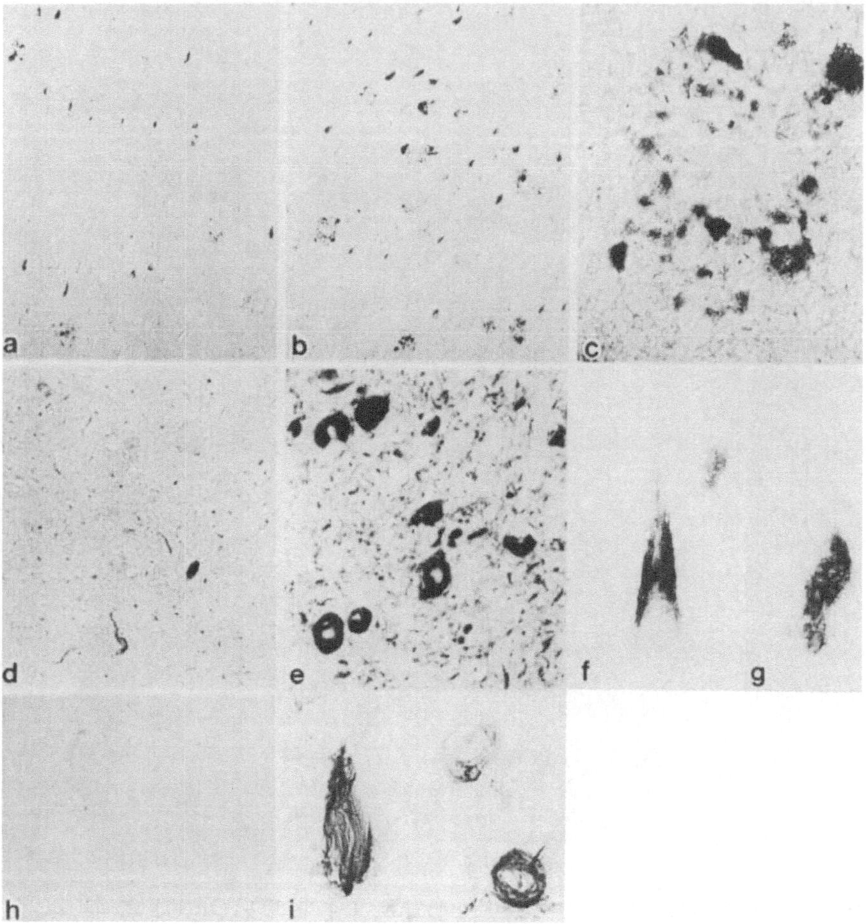

Fig. 3. Immunocytochemical staining with monoclonal antibody to tau and the effect of dephosphorylation. (a–c, f,g). Sections of Alzheimer hippocampus and (d,e) temporal cortex; (h) section of hippocampus of a 80 year old non-Alzheimer dementia individual; (i) tangle-enriched preparation that had been washed twice with 2% (wt/vol) SDS in a boiling water bath. (b,c,e–h) Sections were dephosphorylated with alkaline phosphatase prior to immunolabelling; (a & d) nondephosphorylated controls; adjacent sections and corresponding areas to b & e treated identically except that the alkaline phosphatase was substituted with buffer.
Numbers of immunostained tangles, plaques, and neuropil threads are very much increased in the dephosphorylated tissue sections in b & e as compared to the control treated sections in a & d. (c) Staining of plaque neurites but not of central core amyloid; (f) a neuron with immunolabelled tangle extending into the apical dendrite; (g) a neuron with granulovacuolar inclusions; (h) no staining is seen in the non-Alzheimer hippocampus even after dephosphorylation.
Original magnifications: (a,b) x 75; (c,f,g,i) x 750; (d,e) x 300; (h) x
Reproduced with permission from Grundke-Iqbal, et.al., Proc. Natl. Acad.150.
Sci. U.S.A. 83:4913-17, 1986.

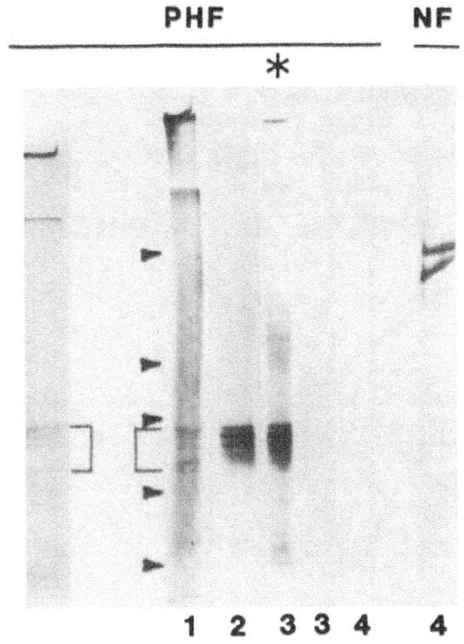

Fig. 4.

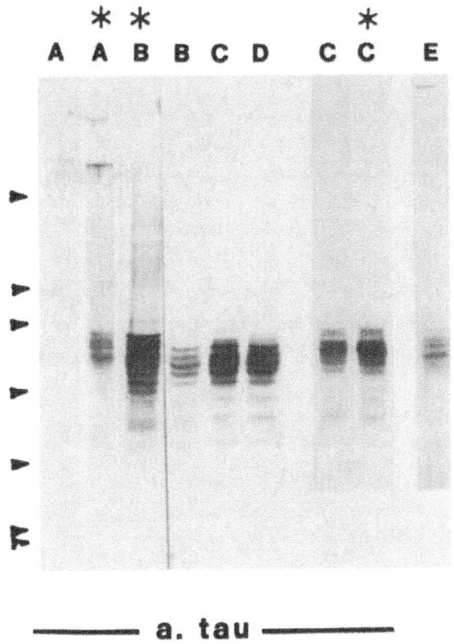

Fig. 5.

Fig. 4. Immunoblots of PHF polypeptides with (lane 1) antiserum to isolated PHF, 1:1000; (lane 2) PHF-reactive anti-microtubule, serum, 1:3000; (lane 3) mAb to tau, 0.1 ug/ml on dephosphorylated (*) and non-dephosphorylated blots and (lanes 4) blots of PHF and neurofilament (NF) polypeptides with mAb to NF, SMI 34, 1:10,000. (*) The dephosphorylation of PHF polypeptides on the paper blots was carried out with alkaline phosphatase (43 ug/ml) before incubation with antibody. Proteins were electro-transferred from SDS-polyacrylamide gel, 5-15% acrylamide gradient. Arrowheads indicate positions of M_r markers from top to bottom: myosin (200,000), phosphorylase b (92,500), bovine serum albumin (68,000), ovalbumin (43,000), α-chymotrypsinogen (25,700). Not shown in this figure, even at a 10-fold increase in the antibody concentration mAb SMI 34 did not label PHF polypeptides. The background smear and the low M_r bands in lane 1 most probably represent oligomers and breakdown products, respectively, of the PHF polypeptides; similar immunostaining pattern is obtained with mAb to PHF.
The far left lane shows the Coomassie blue-stained polypeptide pattern of isolated PHF (5-30% acrylamide gradient).

Fig. 5. Immunoblots with mAb to tau of (lane A) PHF and brain microtubule fractions (lanes B-D) from (lane B) 83 year old Alzheimer patient, and 40 year old (lane C) and 81 year old normal individual (lane D) dephosphorylated (*) or untreated with alkaline phosphatase prior to incubation with the antibody. Proteins were electrotransferred from SDS-polyacrylamide gel, 7-10% acrylamide gradient. Arrowheads indicate positions of M_r markers as in Fig. 2 and in addition β-lactoglobulin (18,400) and cytochrome c (12,300).
The minor tau bands below M_r 45,000 are most likely the degradative products of the 55kDa-62kDa parent tau polypeptides.
Coomassie blue stained patterns of tau purified from calf brain (lane E).
Both Figs. 4 & 5 reproduced with permission from Grundke-Iqbal, et.al. Proc. Natl. Acad. Sci. U.S.A. 83:4913-4917, 1986.

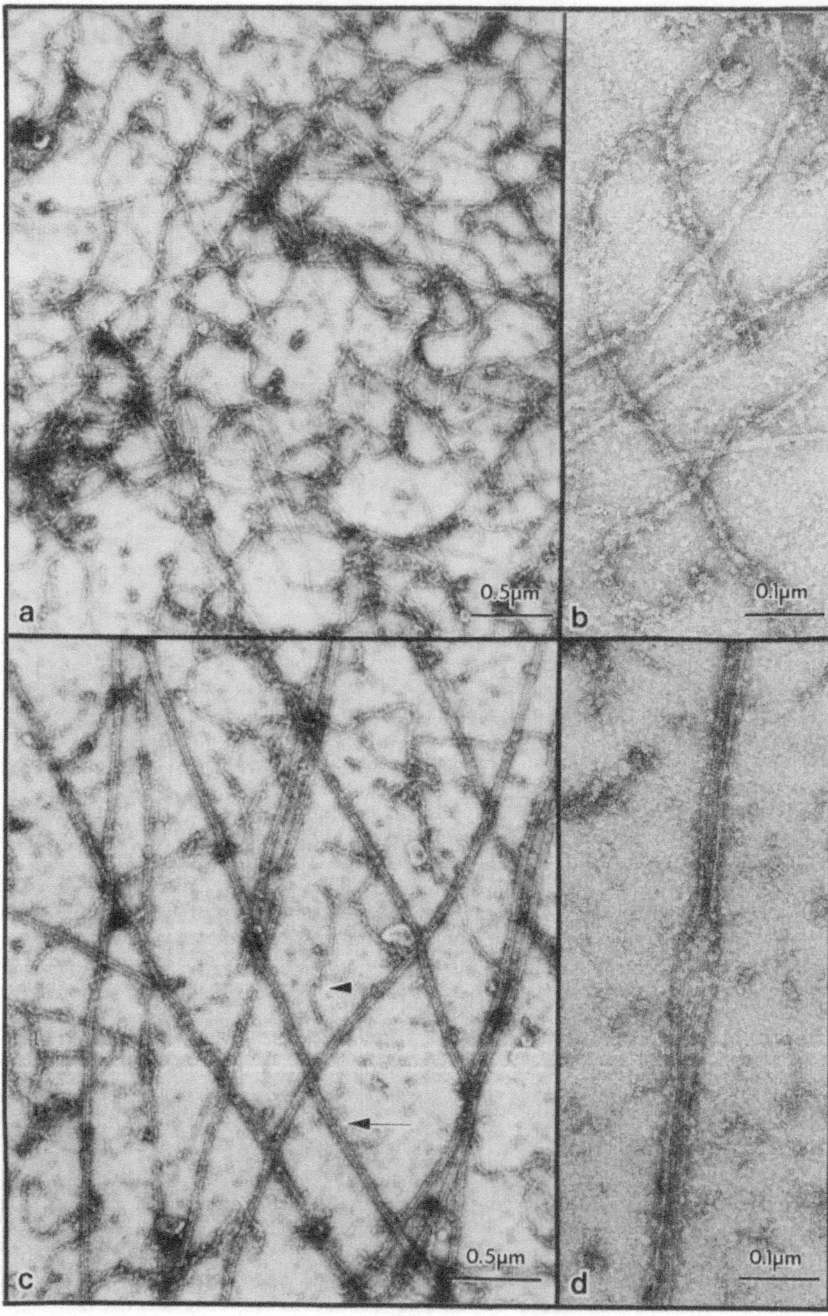

Figure 6: Electron micrographs of in vitro assembled microtubule and filament preparations from a normal control brain stained negatively with 2% phosphotungstic acid, pH 7.2. a and b: Filaments; c: microtubules (←) intermixed with a few 10 nm filaments (◄), from the same control brain as in panel a; a microtubule opened in the middle into a sheet showing protofilaments). Reproduced with permission from Iqbal, et.al., Lancet 2:421-426, 1986.

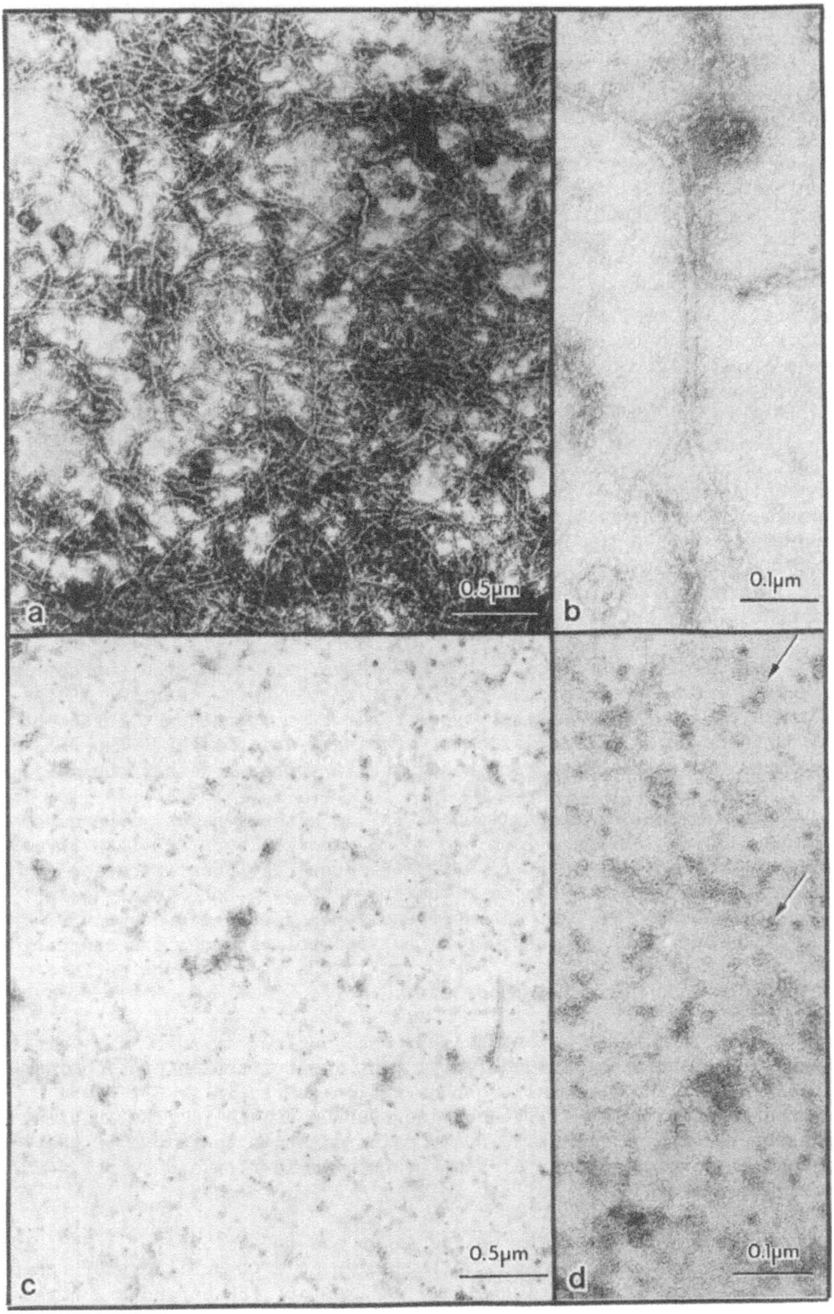

Figure 7: Electron micrographs of in vitro assembled microtubule and filament preparations from an Alzheimer brain stained negatively with 2% phosphotungstic acid, pH 7.2. a and b: filaments; c: microtubule preparation from the same Alzheimer brain as in panel a showing no assembly of microtubules but the presence of numerous rings, 12-14 nm in diameter; d: rings at high magnification; two typical rings are indicated with arrows). Note that diameter of rings is about the same as that of each filament making the PHF. Reproduced with permission from Iqbal, et.al., Lancet 2:422-426, 1986.

SUMMARY

ANT of PHF and neuritic (senile) plaques, which also contain accumulations of PHF in the dystrophic neurites are the most prominent lesions of Alzheimer brain. The presence of these lesions in large numbers determines the histopathological diagnosis of AD/SDAT. Although these lesions were discovered almost as early as the discovery of the disease in 1907 by the German neuropathologist, Alois Alzheimer, most of the studies on the ultrastructure and biochemistry of PHF were reported in the last two decades.

Both ultrastructure and biochemistry of PHF have been controversial. The most widely accepted ultrastructural model for PHF is that it is made up of a pair of 10-13 nm filaments, helically wound around each other. According to one recent report PHF is a twisted ribbon with a left handed-helice.

There are two types of PHF, left handed and right handed PHF. The right handed PHF are larger than the left handed PHF both in diameter and in periodicity. The right handed PHF are readily soluble in SDS and possibly other detergents and denaturants. These two types of PHF might correspond to the two populations of PHF, ANT I and ANT II, readily soluble and insoluble in 2% SDS at room temperature respectively. The ANT II are solubilized when sonicated prior to extraction in SDS and other denaturants.

Methods for bulk isolation of PHF have been developed whereby highly purified preparations of PHF can be obtained. However, these methods employ one or another type of detergent/denaturant treatment which tends to select isolation of sparingly soluble PHF. The PHF isolated under non-denaturing conditions are only modestly enriched.

The polypeptide composition of PHF is heterogeneous. The major PHF polypeptides which range from two to several bands on SDS-PAGE have apparent molecular weights of 45 kDa-62 kDa. The identification of these PHF polypeptides has been confirmed by both monoclonal and polyclonal antibodies to PHF. These PHF polypeptides comigrate on SDS-PAGE and crossreact with microtubule associated polypeptides tau suggesting tau to be a major protein of PHF. Tau in PHF and in Alzheimer brain cytosol is abnormally phosphorylated.

There is a brain microtubule assembly defect in AD/SDAT. This in vitro microtubule assembly defect is not due to tubulin but is most likely due to the abnormal phosphorylation of tau in AD/SDAT brain. Phosphorylation of tau depresses microtubule assembly. The abnormal phosphorylation of tau in AD/SDAT brains might be the cause of the accumulation of PHF by impairment of axoplasmic flow and by polymerization of the abnormal tau in PHF.

Although the major PHF polypeptides and tau comigrate on SDS-PAGE and crossreact immunochemically, the amino acid compositions of tau and PHF are similar (but not identical) and most of PHF accounts for tau immunochemically, the exact chemical relationship between PHF and tau remains to be established. It is not known in addition to tau what are the other components, if any, of PHF. One group of investigators have claimed that PHF like cerebrovascular and plaque amyloid are also made up of the β-protein. However, because of the lack of quantitative data on purity and protein yield this evidence is currently not accepted. It is not clear if the chemical data from this one group of investigators are not the result of the isolation and sequencing of the amyloid protein from the PHF preparations present as a contaminant. Most studies have been unable to detect any crossreactivity between PHF and amyloid either cerebrovascular or plaque core.

Neurofilament is another cytoskeletal element which has been suggested by several groups to be the source of origin of PHF. However, to date only immunocytochemical reactivity of PHF with antibodies to neurofilaments have been shown. All attempts to identify any neurofilament polypeptide by SDS-PAGE or by immunoblots of PHF with antibodies to neurofilaments have been unsuccessful. Furthermore, substructural studies on PHF and neurofilaments have also suggested that PHF are not a product of winding around of neurofilaments. Bodian silver impregnation which stains PHF, neurofilaments and chromatin on tissue sections has been shown to label neurofilament polypeptides separated by SDS-PAGE. However, a silver impregnation method which stains PHF on tissue sections readily stains on SDS-PAGE both PHF and tau polypeptides but not neurofilament triplet. Finally no defect in *in vitro* assembly of neurofilaments from AD/SDAT brain has been detected. The case for the neurofilament origin of PHF with the data available to date is weak.

PHF are also unrelated to scrapie SAF or prion rods in polypeptide composition. To date no crossreactivity between PHF and SAF have been observed.

A final note

Biochemistry of PHF is a very difficult area of research and has been plagued with controversies. Despite it, gaps and differences between different laboratories are getting smaller and smaller everyday. A few years ago there were very few laboratories working on the biochemistry of PHF. Now a number of new several groups continue entering this area of research. A number of difficult problems, namely the bulk isolation of PHF, production of monoclonal and polyclonal antibodies and solubilization of PHF have been achieved. The overcoming of these hurdles and a fantastic increase in interest in this previously neglected area of research is expected to accelerate progress in understanding the primary cause/s of AD/SDAT.

ACKNOWLEDGEMENTS

Studies reviewed in this article from our laboratory were made possible by assistance from Tanweer Zaidi, Yunn-Chyn Tung, Maureen Quinlan, Nasim Ali, Nina Fragola, Sadia S. Shaikh and Lalu George. We would also like to acknowledge the help of summer students, Sabah Zaidi, Ambreen Qureshi, Karen Katz and Cara Della Ventura. Patricia Calimano typed the manuscript.

This work was supported in part by the National Institutes of Health grants AG 05892, NS 18105 and AG/NS 04220.

REFERENCES

Alafuzoff, I. Adolfsson, R. Grundke-Iqbal, I. and Winblad, B. (1985) Perivascular deposits of serum proteins in cerebral cortex in vascular dementia. Acta Neuropathol. (Berl.), 66:292-298.

Allsop, D., Landon, M., Kidd, M., Lowe, J.S., Reynolds, G.P. and Gardener, A. (1986) Monoclonal antibodies raised against a sequence of senile plaque core proteins react with plaque cores, plaque periphery and cerebrovascular amyloid in Alzheimer disease. Neurosci. Lett. 68:252-256.

Anderton, B.H. Breinburg, D. Downes, M.J. Green, P.J. Tomlinson, B.E. Ulrich, J.and Wood, J.N. (1982) Monoclonal antibodies show that neurofibrillary tangles and neurofilaments share antigenic determinants. Nature, 298:84-86.

Bancher, Ch., Lassmann, H., Budka, H., Grundke-Iqbal, I., Iqbal, K., Wiche, G., Seitelberger, F. and Wisniewski, H.M. (1987) Neurofibrillary tangles in Alzheimer's disease (AD) and progressive supranuclear palsy (PSP): antigenic similarities and differences. Microtubule associated protein tau antigenicity is prominent in all types of tangles. Acta Neuropathol. (Berl.) In press.

Bendheim, P.E., Barry, R.A., de Armond, D.J., Stites, D.P. and Prusiner, S.B. (1984) Antibodies to a scrapie prion protein. Nature (London) 310:418-421.

Berl, S. Puszkin, S. and Nicklas, H.J. (1973) Actomyosin-like protein in brain. Science, 179:441-443.

Berzofsky, J.A., Buckenmeyer, G.K., Hicks, G., Gurd, F.R.N., Feldman, R.J., and Minna, J. (1982) Topographic antigenic determinants recognized by monoclonal antibodies to sperm whale myoglobin. J. Biol. Chem. 257:3189-3198.

Braak, H., Braak, E., Grundke-Iqbal, I. and Iqbal, K. (1986) Occurence of neuropil threads in the senile human brain and in Alzheimer's disease: A third location of paired helical filaments outside of neurofibrillary tangles and neuritic plaques. Neurosci. Lett: 65:351-355.

Brion, J.P., Couck, A.M. and Flament-Durand, J. (1984) Ultrastructural study of enriched fractions of tangles from human patients with senile dementia of the Alzheimer type. Acta Neuropathol. (Berl.) 64:148-152.

Brion, J.P. Passareiro, H. Nunez, J. Flament-Durand, J. (1985) Mise en evidence immunologique de la profeine tau an niseau des lesions degenerscence neurofibrillaires de la maladie d'Alzheimer. Arch. Biol (Brux) 95:229-235.

Cleveland, D.W., Hwo, S.Y. and Kirschner, M.W. (1977) Physical and chemical properties of purified tau factor and the role of tau in microtubule assembly. J. Mol. Biol. 116:227-247.

Cork, L.C. Altschuler, R.J. Struble, R.G. Casanova, M.F. Price, D.L. Sternberger, N. and Sternberger, L. (1985) Changes in the distribution of phosphorylated neurofilaments in Alzheimer's disease. J. Neuropathol. Exp. Neurol., 44: Abstr. 183.

Dahl, D. and Bignami, A. (1978) Immunochemical cross-reactivity of normal neurofibrils and aluminum-induced neurofibrillary tangles. Immunofluorescence study with anti-neurofilament serum. Exper. Neurol., 58:74-80.

Dahl, D. Bignami, A. Bich, N.T. and Chi, N.H. (1980) Immunohistochemical characterization of neurofibrillary tangles induced by mitotic spindle inhibitors. Acta Neuropathol. (Berl.) 51:165-168.

Dahl, D. Selkoe, D.J. Pero, R.T. and Bignami, A (1982) Immunostaining of neurofibrillary tangles in Alzheimer's senile dementia with a neurofilament antiserum. J. Neurosci., 2:113-119.

de Garcini, E.M., Serrano, L. and Avila, J. (1986) Self assembly of microtubule associated protein tau into filaments resembling those found in Alzheimer's disease. Biochem. Biophys. Res. Commun. 141:790-796.

Delcourte, A. and Defossez, A. (1986) Alzheimer's disease: tau proteins, the promoting factors of microtubule assembly, are major components of paired helical filaments. J. Neurol. Sci. 76:173-186.

Dentler, W.L. Granett, S. and Rosenbaum, J.L. (1975) Ultrastructural localization of the high molecular weight proteins associated with in vitro assembled brain microtubules. J. Cell Biol., 65:237-241.

Diringer, H. Gelderblom, H. Hilmert, H. Ozel, M. Edelbluth, C and Kimberlin, R.H. (1983) Scrapie infectivity, fibrils and low molecular weight protein. Nature, 306:476-478.

Eng, L.F. Vanderhaegen, J.J. Bignami, A. and Gerstl, B. (1971) An acidic protein isolated from fibrous astrocytes. Brain Res., 28:351-354.
Feit, H. Slusarek, L. and Shelanski, M.L. (1971) Heterogeneity of tubulin subunits. Proc. Nat. Acad. Sci. USA, 68:2028-2031.
Flament-Durand, J., and Brion, J.P. (1986) Neurofibrillary tangles of Alzheimer's disease: ultrastructural and immunocytochemical study. Abst. Intl. Congress Neuropathol., p. 296, Abst. 594.
Gambetti, P. Velasco, M.E. Dahl, D. Bignami, A. Roessmann, U. and Sindely, S.D. (1980) Neurofibrillary tangles in Alzheimer disease; an immunohistochemical study. In: Aging, Vol. 13: Aging of the Brain and Dementia. L. Amaducci, A.N. Davison and P. Antuono (ed.). Raven Press, New York. pp. 55-63.
Gambetti, P., Autilio-Gambetti, L., and Papasozomenos, S.Ch. (1981) Bodian's silver stain method stains neurofilament polypeptides. Science 213:1521-1522.
Gambetti, P. Shecket, G. Ghetti, B. Hirano, A. and Dahl, D. (1983) Neurofibrillary changes in human brain. An immunocytochemical study with a neurofilament antiserum. J. Neuropathol. Exper. Neurol., 42:69-79.
Ghatak, N.R. Nochlin, D. and Hadfield, M.G. (1980) Neurofibrillary pathology in progressive supranuclear palsy. Acta Neuropathol. (Berl.), 52:73-76.
Glenner, G.G. Eanes, E.G. Bladen, H.A. Linke, R.P. and Termine, J.D. (1974) β-pleated sheet fibrils: A comparison of native amyloid with synthetic protein fibrils. J. Histochem Cytochem., 22:1141-1158.
Glenner, G.G. and Wong, C.W. (1984a) Alzheimer's disease: Initial report of the purification and characterization of a novel cerebrovascular amyloid protein. Biochem. Biophys. Res. Commun., 120:855-890.
Glenner, G.G., and Wong, C.W. (1984b) Alzheimer's disease and Down's syndrome: sharing of a unique cerebrovascular amyloid fibril protein. Biochem. Biophys. Res. Commun. 122:1131-1135.
Glenner, G.G. Wong, C.W. Quaranto, V. and Eanes, E.D. (1984) The amyloid deposits in Alzheimer's disease: Their nature and pathogenesis. Appl. Pathol., 2:357-369.
Goldgaber, D., Lerman, M.I., McBride, O.W., Saffiotti, U. and Gajdusek, D.C. (1987) Characterization and chromosomal localization of a cDNA encoding brain amyloid of Alzheimer's disease. Science 235:877-880.
Golub, E.S., and Day, E.D. (1975) Localization of brain-associated hematopoietic antigens in the neuronal fraction of brain. Cell. Immunol. 16:427-431.
Gorevic, P.D., Goni, F., Pons-Estel, B., Alvarez, F., Peress, N.S. and Frangione, B. (1986) Isolation and partial characterization of neurofibrillary tangles and amyloid plaque core in Alzheimer's disease: immunohistological studies. J. Neuropath. Exp. Neurol. 45:647:-664.
Grundke-Iqbal, I. Johnson, A.B. Terry, R.D. Wisniewski, H.M. and Iqbal, K. (1979a) Alzheimer neurofibrillary tangles: Production of antiserum and immunohistological staining. Ann. Neurol., 6:532-537.
Grundke-Iqbal, I. Johnson, A.B. Wisniewski, H.M., Terry, R.D. and Iqbal, K. (1979b) Evidence that Alzheimer neurofibrillary tangels originate from neurotubules. Lancet I:578-580.
Grundke-Iqbal, I., Iqbal, K., Merz, P., and Wisniewski, H.M. (1981) Isolation and properties of Alzheimer neurofibrillary tangles. J. Neuropathol. Exp. Neurol. 40:312 (Abstr.).
Grundke-Iqbal, I., Iqbal, K. and Wisniewski H.M. (1982) Immunocytochemical studies on neurofibrillary changes. Exp. Brain Res. Suppl. 5:20-25.
Grundke-Iqbal, I., Iqbal K. Tung, Y.C. Wang, G.P. and Wisniewski, H.M. (1984) Alzheimer paired helical filaments: Immunochemical identification of polypeptides. Acta Neuropathol. (Berl.), 62:259-267.

Grundke-Iqbal, I., Iqbal, K., Tung, Y.C., Wang, G.P. and Wisniewski, H.M. (1985a) Alzheimer paired helical filaments: Crossreacting polypeptide/s present normally in brain. Acta Neuropathol. (Berl.), 66:52-61.
Grundke-Iqbal, I., Wang, G.P., Iqbal, K., Tung, Y.C. and Wisniewski, H.M. (1985b) Alzheimer paired helical filaments: Identification of polypeptides with monoclonal antibodies. Acta. Neuropathol. (Berl.) 68:279-283.
Grundke-Iqbal, I., Iqbal, K., and Wisniewski, H.M. (1985c) Alzheimer neurofibrillary tangles and plaque neurites crossreact with IgG. J. Neuropathol. Exp. Neurol. 44:368 (Abstr.).
Grundke-Iqbal, I., Iqbal, K., Quinlan, M., Tung, Y-C., Zaidi, M.S., and Wisniewski, H.M. (1986a) Microtubule-associated protein tau: a component of Alzheimer paired helical filaments. J. Biol. Chem. 261:6084-6089.
Grundke-Iqbal, I., Iqbal, K., Tung, Y-C., Quinlan, M., Wisniewski, H.M., and Binder, L.I. (1986b) Abnormal phosphorylation of the microtubule associated protein tau in Alzheimer cytoskeletal pathology. Proc. Natl. Acad. Sci. U.S.A. 83:4913-4917.
Grundke-Iqbal, I., Iqbal, K. and Wisniewski, H.M. (1986c) Microtubule associated protein tau is abnormally phosphorylated in Alzheimer brain. J. Neuropathol. Exp. Neurol. 45:379 (Abstr.).
Grundke-Iqbal, I., Iqbal, K., Wang, G.P. and Wisniewski, H.M. (1987) Microtubule associated protein tau in Alzheimer paired helical filaments (PHF): How much? J. Neuropathol. Exp. Neurol. 46 (Abstr) In press.
Hirano, A., Demlitzer, H.M. and Kurland, L.T. (1968) The fine structure of some intraganglionic alterations. J. Neuropathol. Exp. Neurol. 27:167-182.
Hoffman, P.N. and Lasek, R.J. (1975) The slow component of axonal transport. Identification of major structural polypeptides of the axon and their generality among mammalian neurons. J. Cell Biol., 66:351-356.
Ihara, Y., Abraham, C. and Selkoe, D.J. (1983) Antibodies to paired helical filaments in Alzheimer's disease do not recognize normal brain proteins. Nature, 304:727-730.
Ihara, Y., Nukina, N., Miura, R. and Ogawara, M. (1986) Phosphorylated tau protein is integrated into paired helical filaments in Alzheimer's disease. J. Biochem. (Tokyo) 99:1807-1810.
Iqbal, K., Wisniewski, H.M., Shelanski, M.L., Brostoff, S., Liwnicz, B.H., and Terry, R.D. (1974) Protein changes in senile dementia. Brain Res., 77:337-343.
Iqbal, K. Grundke-Iqbal, I. Wisniewski, H.M. Terry, R.D. (1977a) On neurofilament and neurotubule proteins from human autopsy tissue. J. Neurochem., 29:417-424.
Iqbal, K., Wisniewski, H.M., Grundke-Iqbal, I. and Terry, R.D. (1977b) Neurofibrillary Pathology: An update. In: The Aging Brain and Senile Dementia. K. Nandy and L. Sherwin (Eds.) Plenum, New York, pp. 209-227.
Iqbal, K., Grundke-Iqbal, I., Wisniewski, H.M. and Terry, R.D. (1978) Chemical relationship of the paired helical filaments of Alzheimer's dementia to normal human neurofilaments and neurotubules. Brain Res., 142:321-332.
Iqbal, K., Grundke-Iqbal, I., Johnson, A.B. and Wisniewski, H.M. (1980) Neurofibrous proteins in aging and dementia. In: Aging Vol. 13, Aging of the Brain and Dementia. L. Amaducci, A.N. Davison and P. Antuono (Eds.), Raven Press, New York, pp. 39-48.
Iqbal, K., Merz, P., Grundke-Iqbal, I. and Wisniewski, H.M. (1981) Studies on mammalian neurofilaments isolated by in vitro assembly-disassembly. J. Neuropathol. Exp. Neurol., 40:315 (Abstr.).
Iqbal, K. and Wisniewski, H.M. (1983) Neurofibrillary tangles In:

Alzheimer's Disease. The Standard Reference, B. Reisberg, (Ed.) The Free Press, New York, pp. 48-56.
Iqbal, K., Zaidi, T.. Thompson, C.H., Merz, P.A., and Wisniewski, H.M. (1984) Alzheimer paired helical filaments: Bulk isolation, solubility and protein composition. Acta Neuropathol. (Berl.), 62:167-177.
Iqbal, K., Grundke-Iqbal, I., Zaidi, T., Merz, P.A., Wen, G.Y., Shaikh, S.S., Wisniewski, H.M., Alafuzoff, I., and Winblad, B. (1986a) Defective brain microtubule assembly in Alzheimer's disease. Lancet 2:421-426.
Iqbal, K., Grundke-Iqbal, I., Zaidi, T., Ali, N., and Wisniewski, H.M. (1986b) Are Alzheimer neurofibrillary tangles insoluble polymers? Life Sci. 38:1695-1700.
Iqbal, K., Grundke-Iqbal, I. and Wisniewski, H.M. (1986c) Microtubule associated protein tau is a major constituent of Alzheimer paired helical filaments. J. Neuropathol. Exp. Neurol. 45:378 (Abstr.).
Iqbal, K., Grundke-Iqbal, I. and Wisniewski, H.M. (1987a) Alzheimer's disease, microtubule and neurofilament proteins, and axoplasmic flow. Lancet, 1:102.
Iqbal, K., Grundke-Iqbal, I. and Wisniewski, H.M. (1987b) A silver impregnation method specific for Alzheimer paired helical filaments (PHF) stains microtubule associated protein tau. J. Neruopathol. Exp. Neurol. 46: (Abstr.). In press.
Iqbal, K., Grundke-Iqbal, I. and Wisniewski, H.M. (1987c) Microtubule associated protein tau in Alzheimer paired helical filaments (PHF) J. Neurochem. (Abstr.). In press.
Iqbal, K., Grundke-Iqbal, I., Merz, P.A., Wisniewski, H.M. and Zaidi, T. (1987d) In vitro assembly and isolation of neurofilaments and microtubules from mammalian CNS. Brain Res. In press.
Ishii, T., Haga, S. and Tobutake, S. (1979) Presence of neurofilament protein in Alzheimer's neurofibrillary tangles (ANF); an immunofluorescent study. Acta Neuropathol. (Berl.), 48:105-112.
Kang, J., Lemaire, H.G., Unterbeck, A., Salbaum, J.M., Masters, C.L., Grzeschik, K.H., Multhaup, G., Beyreuther, K. and Muller-Hill, B. (1987) The precursor of Alzheimer's disease amyloid A4 protein resembles a cell-surface receptor. Nature 325:733-736.
Kidd, M. (1963) Paired helical filaments in electron microscopy of Alzheimer's disease. Nature 197:192-193.
Kidd, M. (1964) Alzheimer's disease. An electron microscopical study. Brain, 87:307-320.
Kirschner, C.A., Abraham, C. and Selkoe, D.J. (1986) X-ray diffraction from intraneuronal paired helical filaments and extraneuronal amyloid fibers in Alzheimer's disease indicates cross conformation. Proc. Natl. Acad. Sci. U.S.A. 83:503-507.
Kosik, K.S., Duffy, L.K., Dowling, M.M., Abraham, C., McCluskey, A., and Selkoe, D.J. (1984) Microtubule-associated protein 2: Monoclonal antibodies demonstrate the selective incorporation of certain epitopes into Alzheimer neurofibrillary tangles. Proc. Natl. Acad. Sci. U.S.A. 81:7941-7945.
Kosik, K.S., Joachim, C.L. and Selkoe, D.J. (1986) The microtubule associated protein, tau, is a major antigenic component of paired helical filaments in Alzheimer's disease. Proc. Natl. Acad. Sci. U.S.A. 83:4044-4048.
Lindwall, G., and Cole, R.D. (1984) Phosphorylation affects the ability of tau protein to promote microtubule assembly. J. Biol. Chem. 259:5301-5305.
Masters, C.L., Simms, G., Weinman, N.A., Multhaup, G., McDonald, B.L., and Beyreuther, K. (1985a) Amyloid plaque core protein in Alzheimer disease and Down syndrome. Proc. Natl. Acad. Sci. U.S.A. 82:4245-4249.
Masters, C.L., Multhaup, G., Sims, G., Pottigiesser, J., Martins, R.N., and Beyreuther, K. (1985b) Neuronal origin of a cerebral amyloid:

neurofibrillary tangles of Alzheimer's disease contain the same protein as the amyloid of plaque cores and blood vessels. EMBO J. 4:2757-2763.
Merz, P.A., Sommerville, R.A., Wisniewski, H.M. and Iqbal, K. (1981) Abnormal fibrils from scrapie-infected brain. Acta Neruopathol. (Berl.), 54:63-74.
Merz, P.A., Sommerville, R.A., Wisniewski, H.M., Manuelidis, L. and Manuelidis, E. (1983) Scrapie-associated fibrils in Creutzfeld-Jacob disease. Nature, 306:474-476.
Merz, P.A., Grundke-Iqbal, I., Iqbal, K., Rubenstein, R., and Wisniewski, H.M. (1986) Immune electron microscopy of Alzheimer paired helical filaments (PHF). J. Neuropathol. Exp. Neurol. 45:338 (Abstr.).
Murphy, D.B. and Borisy, G.G. (1975) Association in high molecular weight proteins with microtubules and their role in microtubule assembly in vitro. Proc. Natl. Acad. Sci. USA, 72:2696-2700.
Nigg, E.A., Walter, G., and Singer, S.J. (1982) On the nature of cross-reactions observed with antibodies directed to defined epitopes. Proc. Natl. Acad. Sci. U.S.A. 79:5939-5943.
Nukina, N., and Ihara (1986) One of the antigenic determinants of paired helical filaments is related to tau protein. J. Biochem. (Tokyo) 99:1541-1544.
Olmsted, J.B. and Borisy, G.G. (1973) Microtubules. Ann. Rev. Biochem., 42:507-540.
Perry, G., Rizzuto, N., Autilio-Gambetti, L., and Gambetti, P. (1985) Alzheimer's paired helical filaments contain cytoskeletal components. Proc. Natl. Acad. Sci. U.S.A. 82:4274-4276.
Pollock, N.J., Mirra, S.S., Binder, L.I., Hansen, L.A. and Wood, J.G. (1986) Filamentous aggregates in Pick's disease, progressive supranuclear palsy, and Alzheimer's disease share antigenic determinants with microtubule-associated protein, tau. Lancet 2:1211.
Prusiner, S.B., Bolton, D.C., Groth, D.F., Bowman, K.A., Cochran, S.P. and McKinley, M.P. (1982) Further purification and characterization of scrapie prions. Biochemistry, 21:6942-6950.
Prusiner, S.B., McKinley, M.P., Bowman, K.A., Bolton, D.C., Bendheim, P.E., Groth, D.F. and Glenner, G.G. (1983) Scrapie prions aggregate to form amyloid-like birefringent rods. Cell, 35:349-358.
Robakis, N.K., Wisniewski, H.M., Jenkins, E.C., Devine-Gage, E.A., Houck, G.E., Yao, X.L., Ramakrishna, N., Wolfe, G., Silverman, W.P. and Brown, W.T. (1987) Chromosome 21q21 sublocalization of the gene encoding the beta-amyloid peptide present in cerebral vessels and neuritic (senile) plaques of people with Alzheimer disease and Down syndrome. Lancet 1:384-385.
Roberts, G.W., Crow, T.J., and Polak, J.M. (1985) Location of neuronal tangles in somatostatin neurones in Alzheimer's disease. Nature 314:92-94.
Rubenstein, R., Kascsak, R.J., Merz, P.A., Wisniewski, H.M., Carp, R.I., and Iqbal, K. (1986) Paired helical filaments associated with Alzheimer disease are readily soluble structures. Brain Res. 372:80-88.
Schlaepfer, W.W. (1977) Immunological and ultrastructural studies of neurofilaments isolated from rat peripheral nerve. J. Cell Biol., 74:226-240.
Selkoe, D.J. (1980) Altered protein composition of isolated human cortical neurons in Alzheimer disease. Ann. Neurol. 8:468-478.
Selkoe, D.J., Brown, B.A., Salazar, F.J. and Marotta, C.A. (1981) Myelin basic protein in Alzheimer disease neuronal fractions and mammalian neurofilament preparations. Ann. Neurol. 10:429-436.
Selkoe, D.J., Ihara, Y. and Salazar, F.J. (1982) Alzheimer's disease: Insolubility of partially purified paired helical filaments in sodium dodecyl sulfate and urea. Science, 215 1243-1245.

Selkoe, D.J., Abraham, C.R., Podlinski, M.B., and Duffy, L.K. (1986) Isolation of low-molecular weight proteins from amyloid plaque fibers in Alzheimer's disease. J. Neurochem. 46:1820-1834.
Shihayama, H. and Kitoch, J. (1978) Electron microscopic structure of the Alzheimer's neurofibrillary changes in case of a typical senile dementia. Acta Neuropathol. (Berl.), 48:27-30.
Shirahama, T., Skinner, M., Westermark, P., Rubinow, A., Cohen, A.S., Brun, A.and Kemper, T.L. (1982) Senile cerebral amyloid. Prealbumin as a common constituent in the neuritic plaque, in the neurofibrillary tangle and in the microangiopathic lesion. Am. J. Pathol., 107:41-50.
Soifer, D., Iqbal, K., Czosnek, H., DeMartini, J., Sturman, J.A. and Wisniewski, H.M. (1981) The loss of neuron-specific proteins during the course of Wallerian degeneration of optic and sciatic nerve. J. Neurosci., 1:461-470.
Stam, F.C. and Eikelenboom, P. (1985) Immunopathological study of cerebral senile amyloid. Congophilic angiopathy and senile plaques. In: Interdispl. Topics Geront., Vol. 19, F.C. Rose, (Eds.) Modern Approaches to the Dementias, Part 1:S. Karger, Basel, pp.127-130.
Sternberger, N.H., Sternberger, L.A., and Ulrich, J. (1985) Aberrant neurofilament phosphorylation in Alzheimer disease. Proc. Natl. Acad. Sci. U.S.A. 82:4274-4276.
Tanzi, R.E., Gusella, J.F., Watkins, P.C., Bruns, G.A.P., St. George-Hyslop, P., Van Keuren, M.L., Patterson, D., Pagan, S., Kurmit, D.M. and Neve, R.L. (1987) Amyloid protein gene: cDNA, mRNA distribution, and genetic linkage near the Alzheimer locus. Science, 235:880-884.
Tellez-Nagel, I. and Wisniewski, H.M. (1973) Ultrastructure of neurofibrillary tangles in Steele-Richardson-Olszewski Syndrome. Arch. Neurol., 29:324-327.
Terry, R.D. (1963) The fine structure of neurofibrillary tangles in Alzheimer's disease. J. Neuropathol. & Exper. Neurol., 22:629-642.
Terry, R.D., Gonatas, N.K. and Weiss, M. (1964) Ultrastructural studies in Alzheimer's presenile dementia. Am. J. Pathol,. 44:269-297.
Tomlinson, B.E., Blessed, G. and Roth, M. (1970) Observations on the brains of demented old people. J. Neurol. Sci., 11:205-242.
Wang, G.P., Grundke-Iqbal, I., Kascsak, R.J., Iqbal, K. and Wisniewski, H.M. (1984) Alzheimer neurofibrillary tangles: Monoclonal antibodies to inherent antigen/s. Acta Neuropathol. (Berl.), 62:268-275.
Weingarten, M.D., Lockwood, A.H., Hwo, S.Y.and Kirschner, M.W. (1975) A protein factor essential for microtubule assembly. Proc. Natl. Acad. Sci. USA, 72:1858-1862.
Wen, G.Y. and Wisniewski, H.M. (1984) Substructures of neurofilaments. Acta. Neuropathol. (Berl.), 64:339-343.
Wischik, C.M., Crowther, R.A., Stewart, M., and Roth, M. (1985) Subunit structure of paired helical filaments in Alzheimer's disease. J. Cell. Biol. 100:1905-1912.
Wisniewski, H.M., Narang, H.K. and Terry, R.D. (1976) Neurofibrillary tangles of paired helical filaments. J. Neurol. Sci., 27:173-181.
Wisniewski, H.M., Merz, P.A. and Iqbal, K. (1984) Ultrastructure of paired helical filaments of Alzheimer's neurofibrillary tangle. J. Neuropathol. Exper. Neurol., 43:643-656.
Wisniewski, H.M., Merz, P.A., Kascsak, R.J., Rubenstein, R., Carp, R.I. and Lassmann, H. (1985) Crossreactivity of SAF protein and scrapie amyloid plaques. J. Neuropathol. Exp. Neurol., 44: Abstr. 169.
Wisniewski, H.M. and Wen, G.Y. (1985) Substructures of paired helical filaments from Alzheimer's disease neurofibrillary tangles. Acta Neuropathol. (Berl.), 66:173-176.
Wisniewski, K., Jervis, G.A., Moretz, R.C. and Wisniewski, H.M. (1979) Alzheimer neurofibrillary tangles in diseases other than senile and presenile dementia. Ann. Neurol., 5:288-294.

Wisniewski, H.M., Wen, G.Y., Wang, K.C., Iqbal, K. and Rubenstein, R. (1986) Determination of the handedness of paired helical filaments in Alzheimer's disease. In: Electron Microscopy and Alzheimer's Disease. J. Metuzals (ed.)San Francisco Press, San Francisco, pp.21-24.
Wood, J.G., Mirra, S.S., Pollock, N.J. and Binder, L.I. (1986) Neurofibrillary tangles of Alzheimer's disease share antigenic determinants with the axonal microtubule-associated protein tau. Proc. Natl. Acad. Sci. U.S.A. 83:4040-4043.
Wong, C.W., Quaranta, V., and Glenner, G.G. (1985) Neuritic plaques and cerebrovascular amyloid in Alzheimer's disease are antigenically related. Proc. Natl. Acad. Sci. U.S.A. 82:8729-8732.
Yagishita, S., Itoh, Y. Nan, W. and Amano, N. (1981) Reappraisal of the fine structure of Alzheimer's neurofibrillary tangles. Acta Neuropathol. (Berl.), 54:239-246.
Yan, S-C., Hwang, S., Rustan, T.D., and Frey, W.H. (1985) Human brain tubulin purification: decrease in soluble tubulin with age. Neurochem. Res. 10:1-18.
Yen, S.H., Gaskin, F. and Terry, R.D. (1981) Immunocytochemical studies of neurofibrillary tangles. Am. J. Pathol., 104:77-89.
Yen, S.H., Gaskin, F. and Fu, S.M. (1983) Neurofibrillary tangles in senile dementia of the Alzheimer type share an antigenic determinant with intermediate filaments of the vimentin class. Am. J. Pathol. 113:373-381.
Yen, S.H., Dickinson, D., Crowe, A., Butler, M. and Shelanski, M. (1986) Alzheimer neurofibrillary tangles contain tau proteins. Abstr. Intl. Congress Neuropathol., p. 288, Abstr. 478.

ALTERATION OF THE NEUROFILAMENT-MICROTUBULE NETWORK IN ALZHEIMER DISEASE AND OTHER NEURODEGENERATIVE DISORDERS

George Perry, Valeria Manetto, Michelle Onorato, Madelyn Weiss, Paul Mulvihill, Pamela Galloway, Massimo Tabaton, Lucila Autilio-Gambetti, and Pierluigi Gambetti

Division of Neuropathology, Institute of Pathology, Case Western Reserve University, Cleveland, Ohio

INTRODUCTION

The major lesions consistently found in the brains of Alzheimer patients are neurofibrillary tangles (NFT) and senile plaques (SP). The prevalence of NFT and SP have been found to correlate with the severity of the dementia (1-3). NFT are primarily composed of paired helical filaments (PHF), although 15 nm straight filaments and amorphous components are also common constituents (4). SP, or neuritic plaques, are composed of a core of cerebral amyloid, 6 to 8 nm filaments (5), surrounded by altered neurites containing PHF (6). The chemical identity, mode of formation and physiological consequences of the PHF have been the focus of many studies on Alzheimer disease (AD).

In the first description of AD in 1907, Alois Alzheimer, noting that NFT and neurofibrils are both argentophilic, suggested that NFT result from the accumulation of neurofibrils (7). The modern era of studies on NFT began with Kidd's observation that NFT are composed of PHF, two helically twisted 10 nm filaments (8). This observation was used to support the hypothesis that PHF, and thus NFT, result from an alteration in the components of the neuronal cytoskeleton, with 10 nm neurofilaments being obvious candidates. This chapter will critically review data concerning the origin and composition of these abnormal filaments.

IMMUNOCYTOCHEMISTRY OF NEUROFIBRILLARY TANGLES OF ALZHEIMER DISEASE

Approaches used to identify epitopes in NFT

The reliance of studies of PHF composition on immunocytochemical approaches is a consequence of the chemical properties inherent to PHF: their insolubility (9) and lack of a pure fraction. These factors have impeded the application of direct biochemical analysis to determine PHF composition. Two mirror immunocytochemical approaches have been used to understand NFT composition: the search for antibodies raised to normal proteins that react with NFT and the identification of normal proteins recognized by antibodies raised to PHF.

Antibodies to normal proteins reacting with PHF

The use of specific antibody probes has shed light on the composition of NFT. Immunocytochemical studies indicated that NFT share epitopes with microtubule protein (10) as well as neurofilaments (11,12). These studies were further confirmed through the use of monoclonal antibodies specific to various microtubule (13,14) and neurofilament components (15-17). There is now general agreement that NFT share epitopes with NFH (the high molecular weight neurofilament subunit, often designated NF 200 Kd) and the microtubule associated protein, tau.

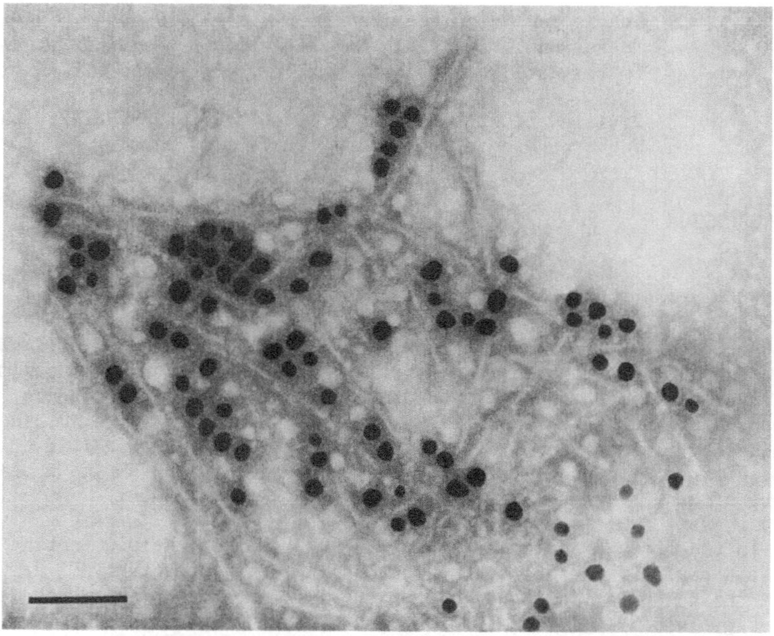

Figure 1. PHF isolated by a procedure incorporating extensive ionic detergent extraction contain cytoskeletal epitopes. The PHF shown were immunodecorated with colloidal gold conjugated to NFH monoclonal antibody 1.1.1. (16) and subsequently negatively stained with 2% phosphotungstic acid. Magnification X 140,000. Scale bar = 0.1 μm.

As discussed in detail in the chapters by Dickson et al, Iqbal et al, Kosik and Neve, Pollock et al, and Selkoe, all antibodies to tau recognize a subpopulation of NFT. This differs from the observations made with antibodies to NFH where only some antibodies that recognize phosphorylated epitopes of NFH are found to recognize NFT (18). Several of these antibodies to NFH which recognize NFT have also been reported to cross react with tau; one of the exceptions is the monoclonal antibody 1.1.1. which does not recognize tau (16,18,19). Therefore, NFT share epitopes with tau and phosphorylated NFH epitopes such as that recognized by monoclonal antibody 1.1.1. Immunochemical methods cannot be used to define whether the shared tau epitopes recognized by other NFH antibodies derive from NFH or tau, or if the epitope in NFT recognized by antibodies like 1.1.1. is an epitope shared between NFH and another as yet unidentified protein in NFT.

Antibodies to PHF recognize normal proteins

PHF have been used as antigen for the production of antibodies useful in the analysis of PHF components. Antibodies to PHF have been demonstrated to react with tau (20,21), NFH (21), ubiquitin (22-25) and with as yet undefined epitopes (21). Although cross-reactivity of antibodies to components of the neuronal cytoskeleton with PHF and cross-reactivity of PHF antibodies with the neuronal cytoskeleton are complementary approaches that yield consistent findings, one must be aware that they both suffer from limitations inherent to immunochemical approaches. Neither approach can be used for the unequivocal demonstration of novel proteins, only novel epitopes. Further, the regions of PHF accessible to antibodies or able to produce an immune response may only be the most exposed or antigenic, not necessarily the most prevalent.

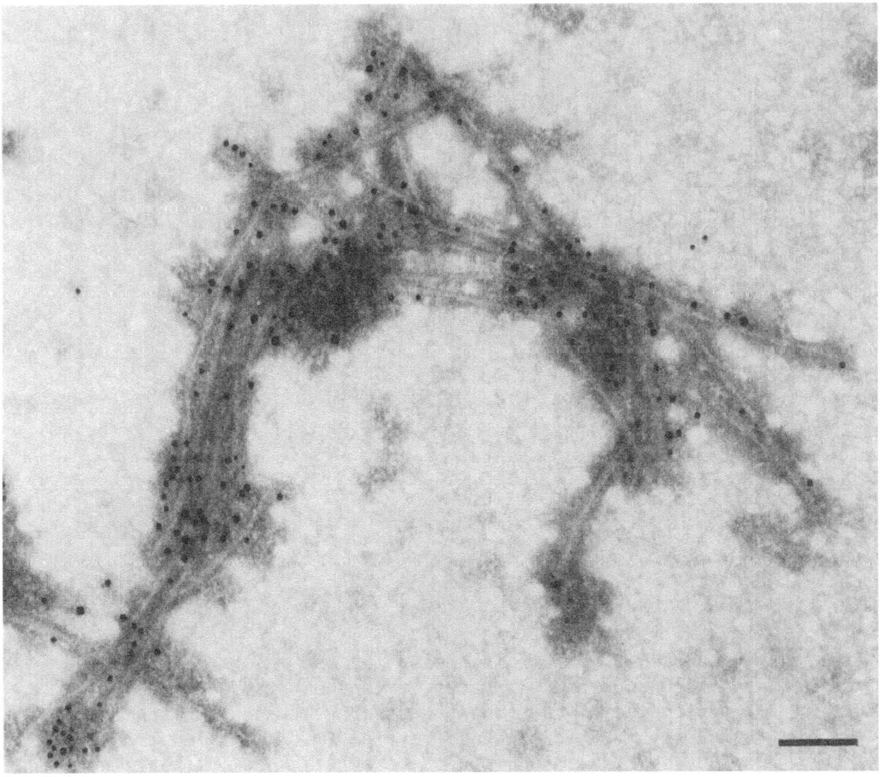

Figure 2. PHF are also decorated by antibodies specifically reacting with tau. These PHF were isolated under conditions of mild denaturation in ionic detergent (26) and were immunodecorated with affinity purified antiserum to tau (27) followed by secondary antibody conjugated to colloidal gold. The preparation was negatively stained with 2% phosphotungstic acid. Magnification X 115,000. Scale bar = 0.1 µm.

As stated above, NFT are primarily composed of PHF but also contain 15 nm straight filaments, amorphous components and occasionally normal cytoskeletal elements (28-30). Immunoelectron microscopy has shown that epitopes recognized by antibodies to tau and NFH are present in 15 nm straight filaments (31) and amorphous elements found within NFT, as well as in PHF (32). The amorphous elements within NFT are present only in tissue obtained at autopsy and not in biopsy specimens (4,6) suggesting that amorphous elements result from postmortem autolysis. Some antibodies to tau react principally with the amorphous components while others react with both filamentous and amorphous components (13,32,33). These findings, along with the observation that the tau epitopes present in PHF can be removed by trypsin treatment (34), suggest that tau in PHF is accessible to protease and therefore is probably located peripheral to the core.

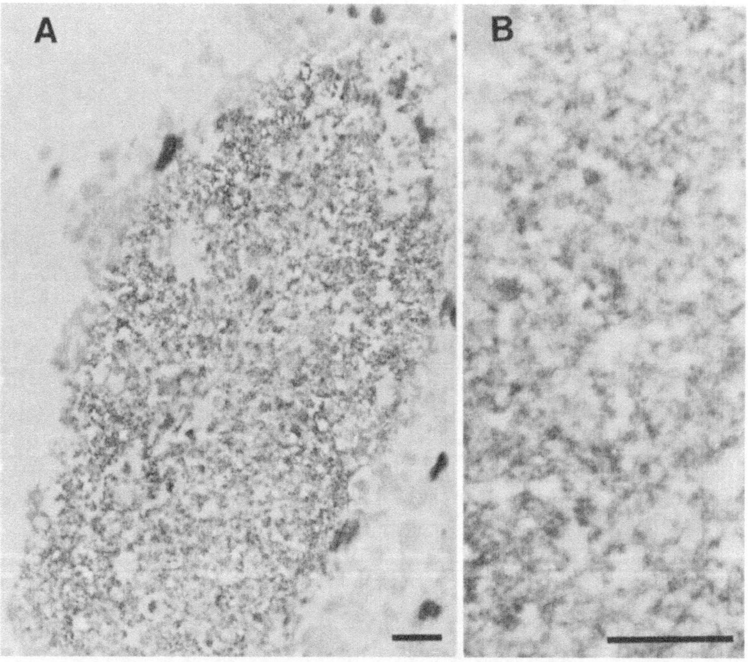

Figure 3. Antibodies to neurofilaments and tau recognize both filamentous and amorphous components within NFT. The antibodies to tau show the strongest reaction with amorphous components. In A, an NFT stained with monoclonal antibody to tau, 5E2 (35), and immunoperoxidase shows the heavy decoration of amorphous components. Even at higher magnification (B), one cannot detect reaction with filaments with this antibody. Magnification A, X 15,000; B, X 37,000. Scale bar = 0.5 μm.

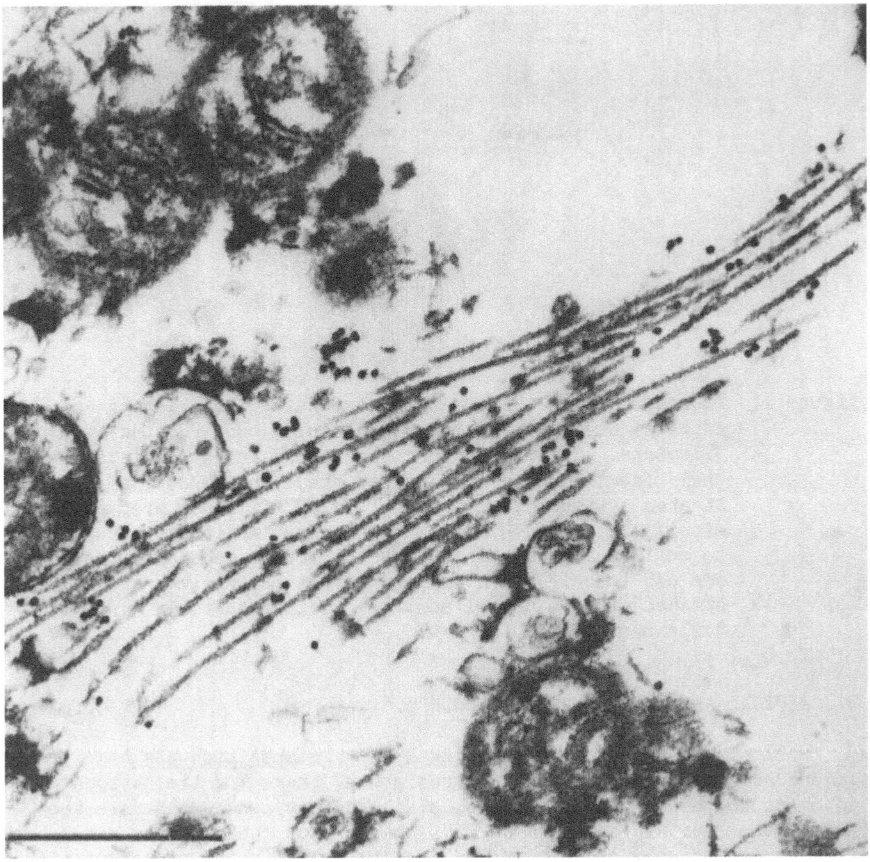

Figure 4. Alzheimer NFT contain both PHF and straight filaments. Straight filaments and PHF share all their identified epitopes. This NFT, containing both filament types, was immunodecorated with an antiserum to PHF (20) followed by secondary antibody conjugated to colloidal gold. Magnification X 62,000. Scale bar = 0.5 μm.

Early morphological studies (28-30,36) have suggested a precursor relationship between straight filaments found in NFT and PHF. The observation that both filament types exist as insoluble polymers (see below) and contain similar microtubule and neurofilament epitopes suggests that they result from a similar transformation of the same cytoskeletal components (31).

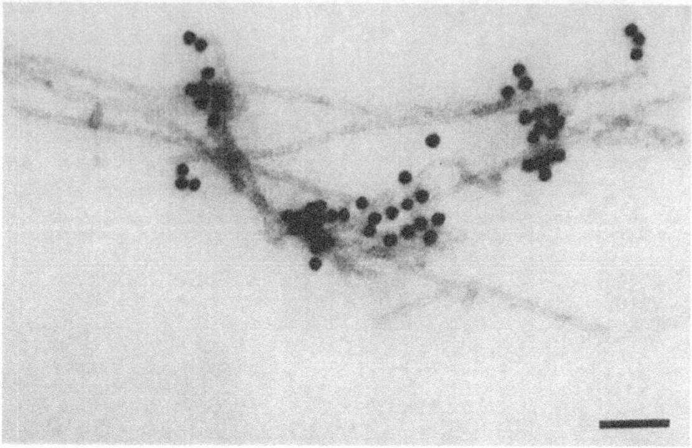

Figure 5: Isolation of NFT filaments by extensive ionic detergent extraction yields fractions containing both PHF and straight filaments. Similar to PHF, straight filaments maintain all of their identified epitopes after these treatments, as demonstrated by immunodecoration of an NFT filament preparation with 1.1.1. (16) followed by secondary antibody conjugated to colloidal gold. In this suboptimally decorated preparation one can identify both filament types as labeled. Negatively stained with 2% uranyl acetate. Magnification X 80,000. Scale bar = 0.1 µm.

BIOCHEMICAL ANALYSIS OF NFT COMPONENTS.

Most of our knowledge concerning NFT/PHF composition has come from immunochemical studies. As discussed above, there are limitations inherent to these methods because, although they can demonstrate the presence of defined epitopes in PHF, they cannot definitively demonstrate the presence of any protein in PHF nor its quantitative representation. Three groups have used biochemical procedures to study PHF composition. A recent elegant biochemical study has demonstrated the presence of ubiquitin sequences and epitopes in PHF (22). Ubiquitin is discussed in another chapter (25). Through extensive sonication and SDS extraction, Iqbal and coworkers have obtained peptides from PHF which can be stained with several antibodies to tau and PHF (14,26,37,38). Although biochemical procedures were used to obtain the peptides, their analysis has thus far depended on immunochemical procedures. Beyreuther and Masters (39) have reported sequence homology between formic acid-solubilized PHF and amyloid the other primary lesion of AD. However, the strong possibility that PHF in these studies were contaminated by amyloid have made this conclusion questionable.

CYTOSKELETAL ELEMENTS PRESENT IN OTHER FILAMENTOUS NEURONAL INCLUSIONS

Determination of the cytoskeletal epitopes found in other filamentous neuronal inclusions has been used to understand NFT formation. The Pick body, the neuronal inclusion of Pick disease, is primarily composed of 12-15 nm straight filaments but also contains twisted filaments similar to PHF (40) and amorphous components. Pick bodies and NFT share all of their identified epitopes (41).

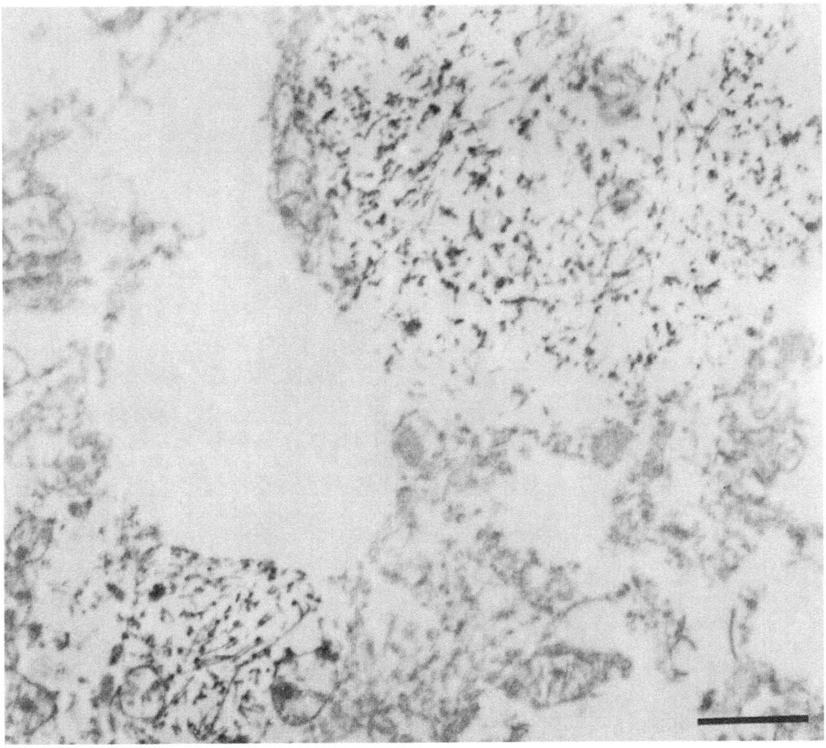

Figure 6. Pick bodies, the characteristic neuronal inclusion found in Pick disease, share all of their identified epitopes with Alzheimer NFT. Monoclonal antibody 1.1.1. (16), recognized filamentous and amorphous components within Pick bodies. Similar results were obtained with other antibodies to NFH, NF, tau and PHF (not shown). Indirect immunoperoxidase. Magnification X 16,000. Scale bar = 1.0 μm.

As was observed for NFT, tau and NFH epitopes in Pick bodies are present in the straight filaments, twisted filaments and amorphous elements. The antibody to PHF, which also recognizes tau and an antisera to microtubule protein reacting with tau, show a more prominent reaction with amorphous components when compared to those to neurofilaments (41). The neurofibrillary tangles found in progressive supranuclear palsy (NFT-PSP) are, like Pick bodies, composed of straight filaments (42,43). The homology between NFT-PSP and Alzheimer NFT is not complete since, although straight filaments of NFT-PSP share epitopes with tau, they do not possess the site recognized by monoclonal antibody 1.1.1. (44). Lewy bodies, found in pigmented neurons of the substantia nigra and locus ceruleus of Parkinson patients, are composed of 10-20 nm straight filaments radiating like a halo from a central amorphous core. Some antibodies to each of the NF triplet proteins (NFL, NFM, NFH) (45), to Alzheimer PHF fractions and tubulin recognize the filamentous halo region of Lewy bodies. Differing from NFT and Pick bodies, Lewy bodies do not contain the epitope recognized by 1.1.1 or share epitopes with tau (46,47).

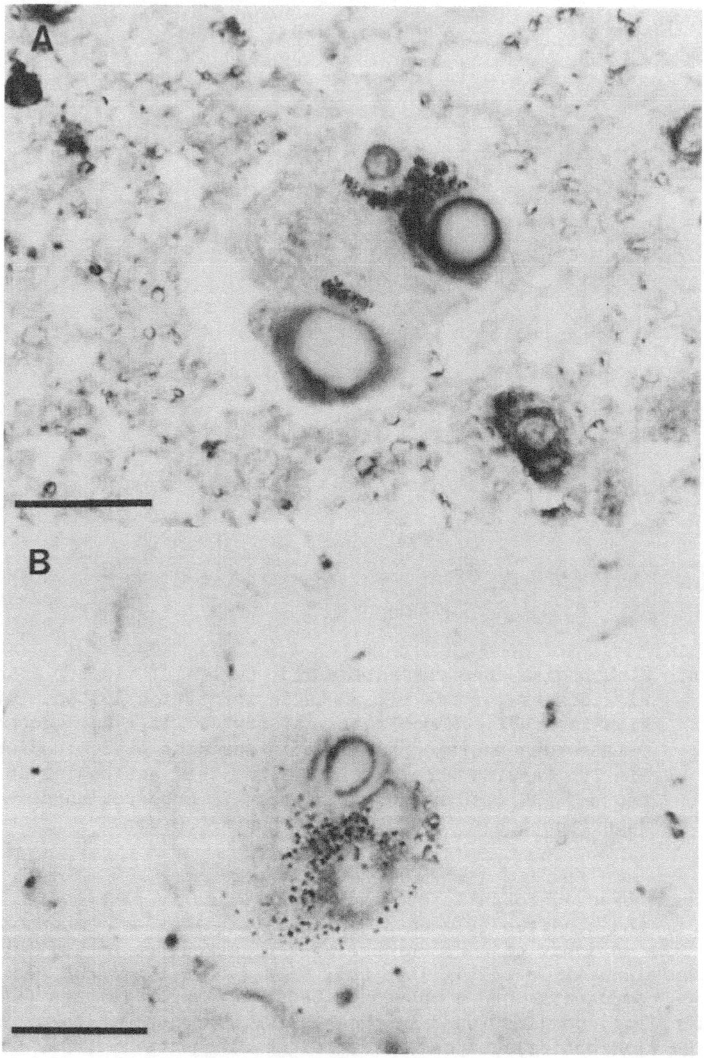

Figure 7. Lewy bodies, the characteristic neuronal inclusion found in selected neurons of Parkinson disease cases, share many of their epitopes with Alzheimer NFT but also contain epitopes not found in NFT. A, Lewy bodies immunostained with an antisera to PHF (48). B, Lewy bodies immunostained with a monoclonal antibody to non-phosphorylated NFH (Sternberger-Meyer Inc, SMI 33). Magnification X 800. Scale bar = 25 μm.

Why do abnormal filaments in these dementing neurodegenerative diseases consistently contain NF and microtubule epitopes while at the same time show antigenic differences? The issue of antigenic similarity in the face of variances in the epitopes found in these inclusions is an important issue basic to understanding the pathogenetic mechanisms responsible for their formation. However, at present we do not know whether to give greater weight to the similarities or to the variances.

INSOLUBILIZATION OF ABNORMAL FILAMENTS

The normal neuronal cytoskeleton (neurofilaments, microtubules and microfilaments) are polymers held together by non-covalent, and in the case of the latter two, dynamic bonds. The cytoskeletal polymers are readily disassociated by denaturants. PHF differ from normal cytoskeletal polymers in that they are insoluble in strong denaturants (9), an observation indicating that bonds maintaining PHF integrity differ from those maintaining cytoskeletal polymers. The epitopes that PHF share with NFH and tau are also insoluble in strong denaturants (32,49) indicating that when these components are incorporated into PHF they are altered from their normal state. The cytoskeletal and other identified epitopes found

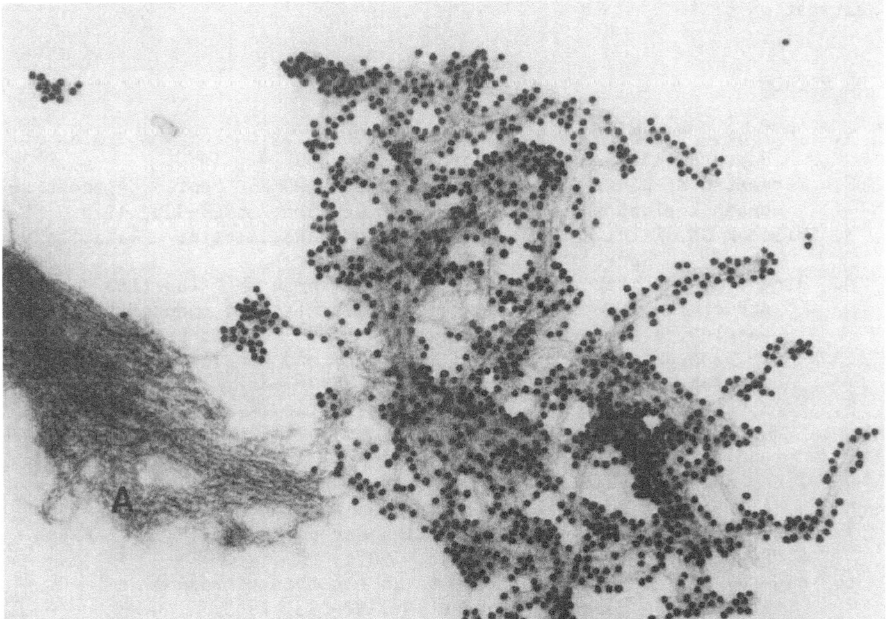

Figure 8. Cytoskeletal epitopes have not been detected in the other filamentous material in Alzheimer cases, cerebral amyloid. An antiserum to neurofilaments (12) strongly decorates PHF but not a bundle of amyloid filaments (A). Magnification X 64,500. Scale bar = 0.5 μm.

in the straight filaments of Pick bodies (41), NFT-PSP (44) as well as NFT of AD (31) are also insoluble. These results indicate that the process of insolubilization is not unique to the Alzheimer PHF.

CONCLUSION AND HYPOTHESES

The relationship between amyloid and PHF is important since AD has been linked to an alteration or duplication of the DNA sequence containing the gene for the amyloid precursor (50-53). Understanding the etiology of AD would be simplified if the two main structural changes of AD could be attributed to the same genetic defect. It has been suggested that both PHF and amyloid are derived from the same protein precursor, a putative integral membrane protein (52). Amyloid would be composed of sequences from the extracellular and membrane spanning domains. It is possible that PHF could derive from the intracellular and membrane spanning domains. The intracellular domains of many membrane receptor proteins have been shown to interact with cytoskeletal elements (54,55); therefore, the presence of intracellular domains in PHF explains the integration of cytoskeletal elements in PHF. This hypothesis can explain the observation that a given NFT can share epitopes with either tau alone, NFH alone, tau and NFH, or neither (56), since these cytoskeletal elements may be specifically, but not invariably, associated with PHF, the core of which would be provided by the amyloid precursor. This hypothesis would explain the presence of the two major structural lesions of AD as the result of the one primary neuronal alteration. The hypotheses engendered by the recent molecular biology studies provide a rationale in understanding PHF and amyloid as they relate to the primary cellular alteration of AD.

REFERENCES

1. Tomlinson BE, Blessed G and Roth M. Observations on the brains of demented old people. J Neurol Sci 11:205-242, 1970.
2. Moramatsu M, Hirai S, Muramatsu A, Yoshikawa M. Senile degeneative brain lesions and dementia. J Am Ger Soc 23:390-406, 1975.
3. Wilcock GK, Esiri MM. Plaques, tangles and dementia: a quantitative study. J Neurol Sci 56:343-356, 1982.
4. Hirano A, Dembitzer HM, Kurland LT, Zimmerman HM. The fine structure of some intraganglionic alterations: neurofibrillary tangles, granulovacular bodies and "rod-like" structures as seen in Guam amyotrophic lateral sclerosis and parkinsonism-dementia complex. J Neuropathol Exp Neurol 27:167-182, 1968.
5. Merz PA, Wisniewski HM, Somerville RA, Bobin SA, Masters CL, Iqbal K. Ultrastructural morphology of amyloid fibrils from neuritic and amyloid plaques. Acta Neuropathol (Berl) 60:113-124, 1983.
6. Terry RD, Gonatas NK, Weiss M. Ultrastructural studies in Alzheimer's presenile dementia. Am J Pathol 44:269-297, 1964.
7. Alzheimer, A. Uber eine eigenartige der Hirnrinde. Allg Z Psych Psych Gerichtl Med 64:146-148, 1907.
8. Kidd M. Paired helical filaments in electron microscopy of Alzheimer's disease. Nature 197:192-193, 1963.
9. Selkoe DJ, Ihara Y, Salazar F. Alzheimer disease: insolubility of purified paired helical filaments in sodium dodecyl sulfate and urea. Science 215:1243-1245, 1982.
10. Grundke-Iqbal I, Johnson AB, Wisniewski HM, Terry, RD, Iqbal K. Evidence that Alzheimer's neurofibrillary tangles originate from neurotubules. Lancet 1:578-580, 1979.
11. Ishii T, Haga S, Tokutake S. Presence of neurofilament protein in Alzheimer's neurofibrillary tangles (ANT). Acta Neuropathol (Berl) 48:105-112, 1979.

12. Gambetti P, Velasco ME, Dahl D, Bignami A, Roessmann U, Sindley SD. Alzheimer neurofibrillary tangles: an immunohistochemical study. In: L Amaducci, AN Davidson, P Antuono (Eds), Aging of the Brain and Dementia, Vol 13, Raven Press, New York,1980, pp 55-63.
13. Wood JG, Mirra SS, Pollock NJ, Binder LI. Neurofibrillary tangles of Alzheimer disease share antigenic determinants with the axonal microtubule-associated protein tau ( ). Proc Natl Acad Sci USA 83:4040-4043, 1986.
14. Grundke-Iqbal I, Iqbal K, Quinlan M, Tung YC, Zaidi MD, Wisniewski, HM. Microtubule-associated protein tau, a component of Alzheimer paired helical filaments. J Biol Chem 261:6084- 6089, 1986.
15. Anderton BH, Breinberg D, Downes MJ, Green PJ, Tomlinson BE, Ulrich J, Wood JN, Kahn J. Monoclonal antibodies show that neurofibrillary tangles and neurofilaments share antigenic determinants. Nature 298:84-86, 1982.
16. Autilio-Gambetti L, Gambetti P, Crane RC. Paired helical filaments: relatedness to neurofilaments shown by silver staining and reactivity with monoclonal antibodies. In: R Katzman (Ed.), Banbury Report 15: Biological Aspects of Alzheimer's Disease, Cold Spring Harbor Lab, New York, 1983, pp 117-125.
17. Sternberger NH, Sternberger LA, Ulrich J. Aberrant neurofilament phosphorylation in Alzheimer disease. Proc Natl Acad Sci USA 82:4274-4276, 1985.
18. Gambetti P, Autilio-Gambetti L, Manetto V, Perry G. Composition of paired helical filaments of Alzheimer disease as determined by specific probes. In: Banbury Report, Davies P. (Ed.), Cold Spring Harbor, New York (in press).
19. Gambetti P, Perry G, Autilio-Gambetti L. Paired helical filaments: do they contain neurofilament epitopes? (Commentary) Neurobiol Aging 7:451-452, 1986.
20. Kosik KS, Joachim CL Selkoe DJ. Microtubule-associated protein is a major antigenic component of paired helical filaments in Alzheimer disease. Proc Natl Acad Sci USA 83:4044-4048, 1986.
21. Perry G, Friedman R, Kang DH, Manetto V, Autilio-Gambetti L, Gambetti P. Antibodies to the neuronal cytoskeleton are elicited by Alzheimer paired helical filament fractions. Brain Res (in press).
22. Mori H, Kondo J, Ihara Y. Ubiquitin is a component of paired helical filaments in Alzheimer's disease. Science 235:1641-1644, 1987.
23. Perry G, Mulvihill P, Manetto V, Fried V, Smith H, Grundke-Iqbal I, Iqbal K, Autilio-Gambetti L, Gambetti P. Ubiquitin in Alzheimer paired helical filaments: a study with monoclonal antibodies. Soc Neurosci. Abstract (submitted).
24. Perry G, Friedman R, Shaw G, Chau V. Ubiquitin is detected in neurofibrillary tangles and senile plaque neurites of Alzheimer disease brains. Proc Natl Acad Sci USA 84:3033-3036, 1987.
25. Perry G, Manetto V, Mulvihill P. Ubiquitin in Alzheimer and other neurodegenerative diseases. In: Alterations of the neuronal cytoskeleton in Alzheimer disease, Perry G (ed.), Plenum Press, New York. (in press)
26. Iqbal K, Zaidi T, Thompson CH, Merz PA, Wisniewski HM. Alzheimer paired helical filaments: bulk isolation, solubility, and protein composition. Acta Neuropathol (Berl) 62:167-177, 1984.
27. Drubin DG, Caput D, Kirschner MW. (1984) Studies on the expression of the microtubule-associated protein, tau, during mouse brain development, with newly isolated complementing DNA probes. J Cell Biol 98:1090-1097.
28. Metuzals J, Montpetit V, Clapin DF. Organization of the neurofilamentous network. Cell Tissue Res 214:455-482, 1981.
29. Metuzals J, Clapin DF, Montpepit V. Formation of paired helical

filaments in Alzheimer's disease. In: 42nd Proc Elec Micr Soc Am, Bailey GW (Ed.) pp 532-533, 1984.

30. Metuzals J. Neurofilaments and paired helical filaments. In: Proc 43rd Ann Meet Elec Micr Soc Am, Bailey GW (Ed.), pp 740-743, 1985.
31. Perry G, Mulvihill P, Manetto V, Autilio-Gambetti L, Gambetti P. Immunocytochemical properties of Alzheimer straight filaments. J Neurosci (in press), 1987.
32. Perry G, Rizzuto N, Autilio-Gambetti L, Gambetti P. Paired helical filaments from Alzheimer disease patients contain cytoskeletal components. Proc Natl Acad Sci USA 82:3916-3920, 1985.
33. Onorato M, Gambetti P, Kosik K, Perry G. Electron microscopic localization of tau in Alzheimer NFT and PHF. J Neuropathol Exp Neurol 46:334, 1987.
34. Merz PA, Wrzolek M, Kasesak RJ, Grundke-Iqbal I, Iqbal K, Binder L, Kosik K, Wisniewski HM. Immunoelectron microscopy of paired helical filaments (PHF) with tau antibodies. J Neuropathol Exp Neurol 46:376, 1987.
35. Galloway PG, Perry G, Kosik KS, Gambetti P. Hirano bodies contain tau protein. Brain Res 403:337-340, 1987.
36. Yoshimura N. Evidence that paired helical filaments originate from neurofilaments. Clin Neuropathol 3:22-27, 1984.
37. Grundke-Iqbal I, Iqbal K, Tung YC, Wisniewski HM. Alzheimer paired helical filaments: identification of polypeptides. Acta Neuropathol (Berl) 62:259-267, 1984.
38. Grundke-Iqbal I, Iqbal K, Tung YC, Quinlan M, Wisniewski HM, Binder LI. Abnormal phosphorylation of the microtubule-associated protein (tau) in Alzheimer cytoskeletal pathology. Proc Natl Acad Sci USA 83:4913-4917, 1986.
39. Masters G, Malthaup G, Simms G, Pottgiesser J, Martins R, Beyreuther K. Neuronal origin of a cerebral amyloid: neurofibrillary tangles of Alzheimer's disease contain the same protein as the amyloid of plaque cores and blood vessels. EMBO J 4:2757-2763, 1985.
40. Munoz-Garcia D, Ludwin SK. Classical and generalized variants of Pick's disease: a clinicopathological, ultrastructural, and immunocytochemical comparative study. Ann Neurol 16:467-480, 1984.
41. Perry G, Stewart D, Manetto V, Autilio-Gambetti L, Gambetti P. Filaments of Pick's bodies contain altered cytoskeletal elements. Am J Pathol 127 (in press).
42. Tellez-Nagel I, Wisniewski HM. Ultrastructure of neurofibrillary tangles in Steele-Richardson-Olsewski syndrome. Arch Neurol 29:324-327, 1973.
43. Bugiani O, Mancardi GL, Brusa A, Ederli A. The fine structure of subcortical neurofibrillary tangles in progressive supranuclear palsy. Acta Neuropathol (Berl) 45:147-152, 1979.
44. Tabaton M, Manetto V, Autilio-Gambetti L, Gambetti P, Fried V, Perry G. Neurofibrillary tangles (NFT) of progressive supranuclear palsy (PSP) and Alzheimer disease (AD): influence of neuronal cell location on antigenic properties. Soc Neurosci Abst (submitted).
45. Goldman JE, Yen SH, Chiu FC, Peress N. Lewy bodies of Parkinson's disease contain neurofilament antigens. Science 221:1082-1084, 1983.
46. Galloway P, Perry G, Grundke-Iqbal I, Autilio-Gambetti L, Gambetti P. Lewy bodies share epitopes with neurofilaments, tubulin and neurofibrillary tangles. Xth Internat Cong Neuropathol 1986, p 298.

47. Galloway PG, Grundke-Iqbal I, Autilio-Gambetti L, Gambetti P, Perry G. Neuronal cytoskeletal involvement in Lewy body formation. J Neuropathol Exp Neurol 46:374, 1987.
48. White CL, Sparkman DR, Gause JM. Distribution of paired helical filaments immunoreactivity in Alzheimer type dementia. J Neuropathol Exp Neurol 44:368, 1985.
49. Miller CCJ, Brion JP, Calvert R, Chin TK, Eagles PAM, Downes MJ, Flament-Durand J, Haugh M, Kahn J, Probst A, Ulrich J, Anderton BH. Alzheimer's paired helical filaments share epitopes with neurofilament side arms. EMBO J 5:269-276, 1986.
50. Goldgaber D, Lerman MI, McBride OW, Saffiotti U, Gajdusek DC. Characterization and chromosomal localization of a cDNA encoding brain amyloid of Alzheimer's disease. Science 235:877-880, 1987.
51. Delabar JM, Goldgaber D, Lamour Y, Nicole A, Huret JL, Grouchy JD, Brown P, Gajdusek DC, Sinet PM. Amyloid gene duplication in Alzheimer's disease and karyotypically normal Down syndrome. Science 235:1390-1392, 1987.
52. Kang J, Lemaire HG, Unterbeck A, Salbaum JM, Masters CL, Grzeschik KH, Malthaup G, Beyreuther K, Muller-Hill B. The precursor of Alzheimer's disease amyloid A4 protein resembles a cell-surface receptor. Nature 325:733-0736, 1987.
53. Tanzi RE, Gusella JF, Watkins PC, Bruns GA, St. George Hyslop P, Kurnit DM, Neve RL. Amyloid beta protein gene: cDNA, mRNA distribution, and genetic linkage near the Alzheimer locus. Science 235:880-884, 1987.
54. Vale R, Shooter E. Alteration of binding properties and cytoskeletal attachment of nerve growth factor receptors in PC12 cells by wheat germ agglutination. J Cell Biol 94:710-717, 1982.
55. Connolly J. Role of cytoskeleton in the formation, stabilization, and removal of acetylcholine receptor clusters in cultured muscle cells. J Cell Bio 99:148-154, 1984.
56. Manetto V, Autilio-Gambetti L, Perry G, Gambetti P. A neurofilament antibody not recognizing tau epitopes stains Alzheimer neurofibrillary tangles (in preparation).

[illegible]

[illegible] Kendall [illegible] Lenox [illegible] Hill [illegible] 1986; [illegible]

[illegible] alterations in [illegible] and chronic [illegible]

[illegible] and humoral [illegible] in rat: [illegible]

[illegible] Shields [illegible] oncogenic [illegible] 1987.

[illegible] 1985; [illegible]

[illegible] of growth [illegible] in vitro [illegible] Science [illegible] 1985.

[illegible] and [illegible] hippocampus [illegible] Brain Research [illegible] 1987.

[illegible] Neuroscience [illegible]

[illegible]

IMMUNOCHEMICAL STUDIES OF THE RELATIONSHIP BETWEEN THE NEURONAL CYTOSKELETON AND NEUROFIBRILLARY TANGLES

D.P.Hue, J.P.Brion, P.A.Robinson, and B.H.Anderton

Department of Immunology
St George's Hospital Medical School
Cranmer Terrace
London SW17 0RE
U.K.

Introduction

Two major types of pathological lesion are present in Alzheimer's disease: the neurofibrillary tangle and the neuritic senile plaque. The presence of tangles in the cerebral cortex has a diagnostic significance in that the frequency of these structures and the extent of choline acetyltransferase loss is correlated with the degree of dementia [1-3]. Tangles are large accumulations of fibrous proteins, the bulk of which has been shown to be composed of paired helical filaments (PHF) [4] which fill the perikarya of affected neurones. These PHF have the appearance of being two 10nm filaments wound around each other to form a helix with a half-periodicity of approximately 80nm [4,5]. Tangles are also localised in many of the abnormal neurites that make up the periphery of the senile plaque [6,7]. The origin of the tangle is unknown although it was suggested that they may be related to neurofilaments [4]. The application of electron microscopy using shadowing and negative staining techniques has revealed the PHF as a left-handed double helix [8] with a relatively short axial length; this makes the structure different from the fibrous sub-structure of neurofilaments [8]. Negative staining of isolated PHF, bovine neurofilaments and microtubules reveals the dissimilarity in visual appearance between these assemblies [9].

However, despite these differences in structure, polyclonal

and monoclonal antibodies raised against neurofilament protein have been shown to recognise epitopes of tangles [10-15]. More recently, reports have also implicated the microtubule-associated protein, tau, in the composition of PHF [16-21]. Antibodies against PHF have been shown to recognise tau protein [16-18] and conversely, anti-tau antibodies have been shown to recognise PHF [18,19]. These reports all suggest that tau is a major component of PHF. In addition, it appears that the tau in Alzheimer brain is abnormally phosphorylated [19,20]. It has been suggested that this phosphorylation of tau may lead to aggregation and formation of insoluble polymers [20] which may in turn interfere with the normal interactions seen between tau and other brain proteins such as tubulin [19,21].

Ihara et al. [10] have shown that antisera raised against the 200K polypeptide of neurofilaments recognises tangles in situ whilst an anti-tubulin antibody failed to do so. Antibodies prepared by Gambetti et al. [12] were shown to react consistently with tangles in situ. More recently, it has been demonstrated that antigenic determinants of neurofilaments and heat-stable microtubule-associated proteins are integral components of PHF [14,15].

There remains uncertainty about the contribution of neurofilaments and microtubule-associated proteins to tangles. Some have argued that PHF originate from amyloid [22-24] and this is based upon the finding that proteins extracted from tangles and amyloid plaque cores are similar. Unfortunately, these reports did not include data describing the degree of purity of the PHF preparations and did not rule out contamination with other insoluble protein residues, including amyloid; our experience is that obtaining pure PHF is technically difficult. PHF have been shown to be a very insoluble structure [25] in an extensive range of chaotrophic buffers and strong protein denaturing reagents [26,27]. This resistance to solubilisation suggests that covalent bonds other than disulphide bonds such as γ-glutamyl - ϵ - lysine cross-linking may be present in PHF [26,28]. This property of insolubility has been exploited by the above authors to obtain partial enrichment of PHF by fractionating brain specimens using strong denaturing conditions.

Disagreement still exists over the degree of insolubility of PHF. Iqbal et al.[29] suggest there are two populations of tangles. One which is readily soluble, the other being solublised by repeated extraction in sodium dodecyl sulphate (SDS) and 2-mercaptoethanol. It has been proposed that as much as 90% of total PHF may be lost during isolation procedures utilising mild SDS treatments. As a result, Rubenstein et al[30] recommended isolation of tangles under non-denaturing conditions since they reasoned that the isolation conditions and not the intrinsic properties of the tangles themselves may be a determining factor for solubility.

We are currently investigating the insolubility of the neurofibrillary tangle and the means of their isolation. By using more stringent strategies we hope to characterise better the protein composition of PHF.

Isolation of PHF

Immunochemical studies have produced two major observations to date: (1) Tangles *in situ* have been stained by some antibodies to neurofilaments and to microtubule-associated proteins[11,14,15,31-33]; (2) Antibodies prepared using preparations enriched in SDS-insoluble PHF as immunogen, stain tangles *in situ* but fail to recognise routinely any normal cellular constituents by immunocytochemistry as possible PHF precursors[34-36].

We recently reported that several neurofilament monoclonal antibodies label SDS-insoluble PHF with one of them, 8D8, labelling more than 90% of the tangle population[15]. Thus, there is evidence to suggest the presence of abnormal cytoskeletal proteins in tangles[11,12,31]. If tangles were simply a mixture of normal neurofilaments and microtubules along with PHF, then all antibodies to these cytoskeletal structures should also immunocytochemically label tangles. However, this is not the case as shown by our studies with neurofilament antibodies[15]. There is a great deal of variation between the proportions of tangles labelled by the antibodies where some label a significant percentage (8D8,RT97) and others none at all (147,RS18). We have, until recently, been isolating tangles utilising a method based upon that of Iqbal et al [29] which has been described in detail previously [15]. In our hands, we find that contamination is a problem. Whilst the method yields an enriched fraction of tangles a large amount of other insoluble material, such as plaque cores and lipofuscin, is also present in our preparations. Various modifications to the method have been tried in order to produce a purer product in larger yield than can easily be achieved by flow cytometry [25,37].

Non-ionic detergents have been widely used in tissue fractionation and they have the advantage of being milder in action than SDS. Hence, if there are two tangle populations [29] then inclusion of SDS during tissue homogenisation might result in a reduced yield of tangles. However, the use of a non-ionic detergent such as Triton X-100 during homogenisation could improve the yield of isolated tangles whilst solubilising membranous contaminants. We have studied the effect of including Triton X-100, Tween 20, Lubrol PX, sodium dodecyl sarcosinate (sarkosyl) and deoxycholate at concentrations of 1% ($^{w}/v$) in the homogenisation buffer. When the final pellet fractions were obtained and stained with thioflavine, the staining showed these preparations to have approximately the same amount of contamination and PHF yield as a preparation where no detergent was used during homogenisation. SDS extraction of the final pellets were made and analysed on a mini SDS-polycrylamide gel electrophoresis system (SDS-PAGE). The results from the gels did not reveal any major differences between these samples (data not shown).

After the homogenate has been prepared in 0.32M sucrose and the sucrose concentrations re-adjusted to 1M the final volume of homogenate is approximately 60ml. 15ml of homogenate is then layered upon a sucrose step-gradient consisting of 20ml 1.2M sucrose over 5 ml 1.8M sucrose [15]. This is similar to the original method where 40 ml of homogenate is layered over a 20 ml discontinuous density gradient [29]. A more efficient separation will be achieved when the volume of homogenate is reduced to a level such that it is less than or equal to the volume of the 1.8M sucrose layer.

We adopted this modification and homogenised 1g of tisssue in 5.5 ml of 0.32M sucrose. The homogenate was filtered once through a 60 µm mesh, admixed with 4ml 2.4 M sucrose and 5 ml of homogenate was layered on each step-gradient. Centrifugation was carried out as normal. We have noted the following:
(1) the population of the tangles at the interface between the 1.2 M and 1.8 M sucrose layer are different to those in the pellet. The tangles at the interface are very fine and threadlike in structure when examined by thioflavine staining, whereas those in the pellet are much larger and consist mainly of whole intact tangles (see figures 1 and 2). (2) The amount of tissue required to isolate these fractions is reduced. Whereas previously, we were routinely using 8g of grey matter, the 1g of tissue used in the modified method appears to have produced a much greater yield of tangles. We have quantified the number of tangles in each preparation by using thioflavine and assessing the number of tangles in a representative field of view. As figures 1 and 2 demonstrate there are many tangles in the modified PHF preparation. By the old method we would obtain around 20% of these yields in a good preparation. Obviously the initial concentration of tangles in the brain is important as well, but these observations have been made from similar areas of the same brain, hence one can consider the initial tangle concentrations to be comparable. We are currently examining these two populations of tangles for differences in properties.

Solubilization of PHF

Masters et al [24] reported that after pepsin treatment PHF were soluble in formic acid and that a polypeptide of 4000 molecular weight was obtained; PHF have been shown to be susceptible, to a degree, to pepsin and other proteases [38,39]. Selkoe [27] has reported that PHF are also insoluble in a phenol: acetic acid: water mixture in addition to the normal chaotrophic reagents [26]. Phenol: acetic acid: water is a strong protein solubilizing medium which has been used previously in the separation of ribosomal proteins and also encephalomyocarditis virus [40]. Samples solubilised in phenol: acetic acid: water have been applied for separation directly to acetic acid: urea polyacrylamide electrophoresis gels (7.5% acrylamide, 35% acetic acid, 5M urea) [41]. We have modified this solvent system and

Figure 1: Thioflavine staining of isolated tangles, harvested at the 1.2M / 1.8M sucrose interface, from a PHF preparation following SDS- extraction.

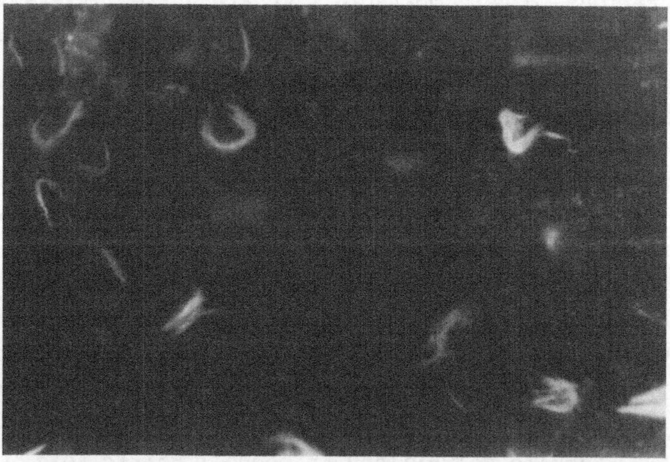

Figure 2: Thioflavine staining of isolated tangles collected in the pellet fraction of a PHF preparation following SDS- extraction.

investigated the ability of phenol: acetic acid: urea (PAU) in a ratio of 2:1:1 (w/v/v) as a solubilizing medium where the urea concentration adopted was 8M.

Tangle fractions were obtained using the method of Miller et al [15] and were either treated immediately with PAU overnight at room temperature (ie, pre-SDS extraction) or after first extracting in SDS. The acetic acid: urea (AU) PAGE system described by Panyim and Chalkley [42] consisting of 10% acrylamide, 0.9M acetic acid and 6.25M urea was used for separations of these samples immediately after solubilization. After staining of the the gels large amounts of streaking was evident possibly due to the presence of phenol in the samples. In order to overcome this effect, the samples solubilized in PAU were dialysed before electrophoresis into 20 % (v/v) acetic acid. During this stage a precipitate formed suggesting that phenol is intrinsically involved in solubilizing these preparations. This precipitate was highly insoluble and did not redissolve in PAU.

When a non-SDS extracted sample was dissolved in PAU and run on AU-PAGE a series of bands was observed with a large amount of excluded material remaining trapped at the top of the gel. When the pellets were first SDS-extracted then solubilized in PAU these lower bands vanished leaving only excluded material at the top of the gel (data not shown). When these gels were immunoblotted using monoclonal antibodies to neurofilaments and a polyclonal to PHF a similar set of results was obtained. Bands were detected on the blot when the samples were not pre-SDS extracted before PAU solubilization, as well as a band corresponding to the excluded material at the top of gel. This latter band was detected by both the neurofilament monoclonal antibodies and the anti-PHF polyclonal antiserum. When the samples were pre-SDS extracted the only band detected was that corresponding to the excluded material and this was labelled by the neurofilament monoclonal antibodies and anti-PHF polyclonal antiserum (see figures 3 and 4). These results show that the bands seen before SDS extraction include normal neurofilament material. After SDS-solubilization these proteins are removed leaving behind the material that is unable to enter the gel. This material is unlikely to contain normal neurofilaments as

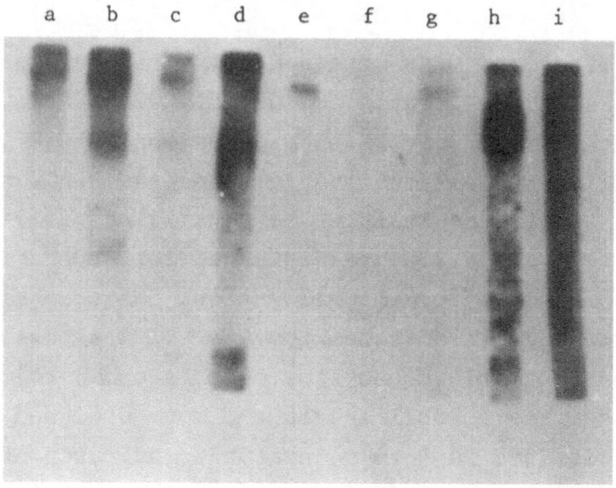

Figure 3: Immunoblot of proteins from a PHF-enriched fraction, before SDS extraction, solubilised in PAU and separated by AU-PAGE. Labelled with a)8D8 b)RT97 c)1215 d)BF10 e)147 f)155 g)RS18 h)anti-IFA and i)anti-PHF.

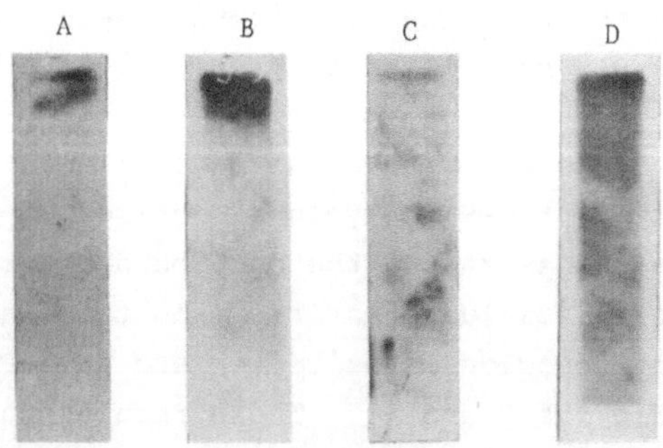

Figure 4: Immunoblot of proteins from a PHF-enriched fraction, after SDS-extraction, solubilised in PAU and separated by AU-PAGE. Labelled with A)8D8 B)RT97 C)anti-IFA and D)anti-PHF.

these will have been removed by SDS extraction. Two possible explanations for this may be: (1) the neurofilaments have been modified in some manner so promoting their aggregation or (2) neurofilaments (peptides or fragments) have been incorporated into tangle structure along with other unknown proteins or polypeptides.

Isolation of cDNA for the 200K neurofilament polypeptide

We have recently reported the isolation of 2 cDNA probes for the 200K polypeptide (NF-H) of neurofilaments [43]. One of these probes (2.5kb) encoded for the C-terminal tail, whilst the other (3.0kb) overlapped and also encoded the epitope for the monoclonal antibody, anti-IFA. The epitope for anti-IFA is believed to be present in all intermediate filaments, located at the highly conserved C-terminal end of alpha-helix 2 in the general structure of intermediate filaments [44]. We belive that the 5' end of the 2.5 kb cDNA corresponds to the NF-H C-terminal tail which is known to be heavily phosphorylated [45]. Our sequence data has revealed this region to contain a peptide repeat unit of -Glu.(Ala, Val).Lys.Ser.Pro.Ala- that is repeated several times. The alanine-valine interchange represents a conservative difference with alanine being predominant. The presence of the serine residue in the centre of a charged environment may represent a phosphorylation site and thus the repeat sequence may form the known multiphosphorylation site in NF-H. The importance of the multiphosphorylation site is unknown although one possibility is that the phosphorylation may be involved in mediating interactions between neurofilaments and other neuronal components [45].

The question of phosphorylation state may have a pivotal role in the formulation of the neurofibrillary tangle. It has been demonstrated that antibodies to phosphorylated neurofilaments stain tangles and plaques in situ but fail to do so after these structures have been pre-treated with phosphatase [46,47]. This suggests that there may be an aberrant phosphorylation of the neurofilaments in the perikaryon before transport to their axonal location can take place. The presence of structurally abnormal phosphorylated neurofilaments in neurofibrillary tangles has been supported by others [47-49]. It has been proposed that the aberrant phosphorylation may cause the neurofilaments to become more compact which prevents their migration into the axon [46] and in addition only one or few of the neurofilament phosphorylation sites is involved. It is now of interest to see if PHF contain this same sequence and these cDNA clones will enable us to map these epitopes on neurofilaments that are either associated with or absent from PHF.

Acknowledgement
This work was supported by The Wellcome Trust and the Medical Research Council.

REFERENCES

1. Farmer PM, Peck A, Terry D: Correlations among numbers of neuritic plaques, neurofibrillary tangles and the severity of senile dementia. J Neuropathol Exp Neurol 35: 367 (1976).
2. Tomlinson BE: The pathology of dementia, in Wells CG (ed): Dementia ed2. Philadephia, FA Davies Co, 1977, p113.
3. Wilcock GK, Esiri MM, Bowen DM, et al: Alzheimer disease: correlation of cortical choline acetyltransferase activity with the severity of dementia and histological abnormalities. J Neurol Sci 57: 407-1982.
4. Kidd M: Paired helical filaments in electron microscopy of Alzheimer's disease. Nature (Lond) 197: 192-193, 1963.
5. Wisniewski HM, Narang NK, Terry RD: Neurofibrillary tangles of paired helical filaments. J Neurol Sci 27: 173-181, 1976.
6. Kidd M: Alzheimer's disease - an electron microscopical study. Brain 87: 307-320, 1964.
7. Gonatas NK, Anderson W, Evangelista I: The contribution of altered synapses in the senile plaque: an electron microscopic study in Alzheimer's dementia. J Neuropathol Exp Neurol 26: 25, 1967.
8. Wischik CM, Crowther RA, Stewart M et al: Subunit structure of paired helical filaments in Alzheimer's disease. J Cell Biol 100: 1905 - 1912, 1985.
9. Miller C, Haugh M, Kahn J, et al: The cytoskeleton and neurofibrillary tangles in Alzheimer's disease. TINS 9: 76-81, 1986.
10. Ihara Y, Nukina N, Sugita H, et al : Staining of Alzheimer's neurofibrillary tangles with antiserum against 200K component of neurofilament. Proc Japan Acad 57: 152-156, 1981.
11. Anderton BH, Breinburg D, Downes MJ, et al : Monoclonal antibodies show that neurofibrillary tangles and neurofilaments share antigenic determinants. Nature (Lond) 298: 84-86, 1982.

12. Gambetti P, Autilio-Gambetti L, Perry G, et al: Antibodies to neurofibrillary tangles of Alzheimer's disease raised from human and animal neurofilament fractions. Lab Invest 49: 430 - 435, 1983.
13. Rasool CG, Selkoe DJ: Alzheimer's disease: exposure of neurofilament immunoreactivity in SDS-insoluble paired helical filaments. Brain Res 322: 194 - 198, 1984.
14. Perry G, Rizzuto N, Autilio-Gambetti L, et al: Paired helical filaments from Alzheimer's disease patients contain cytoskeletal components. Proc Natl Acad Sci (USA) 82: 3916-3920, 1985.
15. Miller CCJ, Brion JP, Calvert R, et al: Alzheimer's paired helical filaments share epitopes with neurofilament side arms. EMBO J 5: 269 - 276, 1986.
16. Nukina N, Ihara Y: One of the antigenic determinants of paired helical filaments is related to tau protein. J Biochem 99: 1541 - 1544, 1986.
17. Grundke-Iqbal I, Iqbal K, Quinlan M, et al: Microtubule-associated protein tau. J Biol Chem 261: 6084-6089, 1986.
18. Kosik KS, Joachim CL, Selkoe DJ: Microtubule-associated protein tau (τ) is a major antigenic component of paired helical filaments in Alzheimer disease. Proc Natl Acad Sci (USA) 83: 4044-4048, 1986.
19. Grundke-Iqbal I, Iqbal K, Tung YC, et al: Abnormal phosphorylation of the microtubule - associated proteins (tau) in Alzheimer cytoskeletal pathology. Proc Natl Acad Sci (USA) 83: 4913 - 4917, 1986.
20. Ihara Y, Nukina N, Miura R, et al: Phosphorylated tau protein is integrated into paired helical filaments in Alzheimer's disease. J Biochem 99: 1807-1810, 1986.
21. Iqbal K, Grundke-Iqbal I, Zaidi T, et al: Defective brain microtubule assembly in Alzheimer's disease. Lancet (ii): 421-426, 1986.
22. Allsop D, Landon M, Kidd M: The isolation and amino acid composition of senile plaque core protein. Brain Res 259: 348 - 352, 1983.
23. Kidd M, Allsop D, Landon M : Senile plaque amyloid, paired helical filaments, and cerebrovascular amyloid in Alzheimer's disease are all deposits of the same protein. Lancet (i): 278, 1985.
24. Masters CL, Multhaup G, Simms G, et al: Neuronal origin of a cerebral amyloid: neurofibrillary tangles of Alzheimer's disease contain the same protein as the amyloid of plaque cores and blood vessels. EMBO J 4: 2757 - 2763, 1985.
25. Miller CJ, Haugh M, Anderton BH, et al: Neurofibrillary tangles from Alzheimer's disease brain purified using a cell sorter. Neurosci Letts 63: 247 - 252, 1986.
26. Selkoe DJ, Ihara Y, Salazar FJ: Alzheimer's disease: Insolubility of partially purified paired helical filaments in sodium dodecyl sulphate and urea. Science 215: 1243-1245, 1982.
27. Selkoe DJ, Ihara Y, Salazar FJ : Biochemical and immunocytochemical studies of Alzheimer paired helical filaments, in Katzman R (ed): Biological aspects of Alzheimer's disease, Banbury Report 15. New York, Cold Spring Harbour Laboratory Press, 1983, p125.

28. Miller CCJ, Anderton BH: Translutaminase and the neuronal cytoskeleton in ALzheimer's disease. J Neurochem 46: 1912 - 1922, 1986.
29. Iqbal K, Zaidi T, Thompson CH, et al: Alzheimer paired helical filaments: bulk isolation, solubility and protein composition. Acta Neuropathol (Berl) 62: 167-177, 1984.
30. Rubenstein R, Kascsak RJ, Merz PA, et al: Paired helical filaments associated with Alzheimer's disease are readily soluble structures. Brain Res 372: 80-88, 1986.
31. Kosik KS, Duffy LK, Dowling MM, et al: Microtubule - associated protein 2: Monoclonal antibodies demonstrate the selective incorporation of certain epitopes into Alzheimer neurofibrillary tangles. Proc Natl Acad Sci (USA) 81: 7941 - 7945, 1984.
32. Grundke-Iqbal I, Iqbal K, Tung YC, et al: Alzheimer paired helical filaments: Cross-reacting polypeptide(s) normally present in brain. Acta Neuropathol (Berl) 66: 52-61, 1985.
33. Brion JP, van den Bosch de Aquilar P, Flament-Durand J,in Traber J, Gispen WH (eds): Advances in applied neurological sciences, vol 2, Senile Dementia of the Alzheimer type. Berlin/Heidelburg, Springer-Verlag, 1985, p164.
34. Ihara Y, Abraham C, Selkoe DJ: Antibodies to paired helical filaments in Alzheimer's disease do not recognise normal brain proteins. Nature (Lond) 304: 727-730, 1983.
35. Grundke-Iqbal I, Iqbal K, Tung YC, et al: Alzheimer paired helical filaments: immunochemical identification of polypeptides. Acta Neuropathol (Berl) 62: 259-267, 1984.
36. Brion JP, Couck AM, Passareiro E, et al: Neurofibrillary tangles of Alzheimer's disease: an immunohistochemical study. J Submicrosc Cytol 17: 89-96, 1985.
37. Hussey S, Yates CM, Fink G: Fluorescence activated cell sorting (FACS) as a separation method for neurofibrillary tangles in Alzheimer's disease. J Neurosci methods 16: 1-8, 1986.
38. Yen SH, Kress Y: The effect of chemical reagents or proteases on the ultrastructure of paired helical filaments, in Katzman R (ed): Biological aspects of Alzheimer's disease, Banbury Report 15. New York, Cold Spring Harbour Laboratory Press,1983, p155.
39. Nukina N, Ihara Y: Proteolytic fragments of Alzheimer's paired helical filaments. J Biochem 98: 1715-1718, 1985.
40. Work TS: Electrophoretic separation of the proteins of ribosome subunits and of encephalomyocarditis virus. J Mol Biol 10:544-544, 1964.
41. Takayama K, MacLennan DH, Tzagoloff A, et al: Studies on the electron transfer system: LXVII polyacrylamide gel electrophoresis of the mitochondrial electron transfer complexes. Arch Biochem Biophys 114: 223-230, 1964.
42. Panyim S, Chalkley R: High resolution acrylamide gel electrophoresis of histones. Arch Biochem Biophys 130: 337 - 246, 1969.

43. Robinson PA, Wion D, Anderton BH: Isolation of a cDNA for the rat heavy neurofilament polypeptide (NF-H). FEBS Lett 209: 203-205, 1986.
44. Geisler N, Kaufmann E, Fischer S, et al: Neurofilament architecture combines structural principles of intermediate filaments with carboxy terminal extensions increasing in size between triplet proteins. EMBO J 2: 1295 - 1302, 1983.
45. Julien JP, Mushynski WE: The distribution of phosphorylation sites among identified proteolytic fragments of mammalian neurofilaments. J Biol Chem 258: 4019 - 4025, 1983.
46. Sternberger NH, Sternberger LA, Ulrich J: Aberrant neurofilament phosphorylation in Alzheimer's disease. Proc Natl Acad Sci (USA) 82: 4274 - 4276, 1985.
47. Haugh MC, Probst A, Ulrich J, et al: Alzheimer neurofibrillary tangles contain phosphorylated and hidden neurofilament epitopes. J Neurol Neurosurg Psych 49: 1213 - 1220, 1986.
48. Bizzi A, Gambetti P: Phosphorylation of neurofilaments is altered in aluminium intoxication. Acta Neuropathol (Berl) 71: 154 - 158, 1986.
49. Cork LC, Sternberger NH, Sternberger LA, et al: Phosphorylated neurofilament antigens in neurofibrillary tangles in Alzheimer's disease. J Neuropathol Exp Neurol 45: 56-64, 1986.

MONOCLONAL ANTIBODIES SHOW CROSS-REACTIVITY OF ALZHEIMER NEUROFIBRILLARY TANGLES AND HEAT-STABLE MICROTUBULE-ASSOCIATED PROTEINS

Dennis W. Dickson, Hanna Ksiezak-Reding,
Alex Crowe, and Shu-Hui Yen.

Department of Pathology and The Rose F. Kennedy Center For
Research in Mental Retardation and Human Development
Albert Einstein College of Medicine, Bronx, New York

INTRODUCTION

Recent biochemical and immunocytochemical studies have provided evidence that heat-stable, microtubule-associated proteins (MAPs) may be instrumental in the pathogenesis of Alzheimer neurofibrillary tangles (ANT)[1-5]. Some investigators speculate that the mid-molecular weight, heat-stable MAP known as tau protein[6] may be the subunit of the paired helical filament (PHF)[2,3], which is the major ultrastructural component of ANT[7]. Tau is a group of proteins with molecular weight of 42-68 kd (Fig. 1) that co-purifies stoichiometrically with tubulin[8]. Despite its microheterogeneity in terms of isoelectric point and molecular weight, all tau proteins have conserved amino acid sequences that are recognized by several antibodies[9]. The microheterogeneity of tau protein may be related to several processes including alternative gene splicing[10] or post-translational modification such as proteolysis[9] or phosphorylation[11,12]. The function of tau proteins, other than facilitating tubulin polymerization into microtubules[8,13], is currently unkown. The distribution of the various isoforms of tau with respect to neuronal type and neuronal subcellular location is incompletely understood at present. Most tau protein appears to be localized to axons, but some is also present in perikarya and dendrites[14]. The role of specific isoforms of tau in ANT remains to be investigated.

Phosphorylation of MAPs may be a mechanism of regulating their functional attributes[15,16], one of which may be the bridging of intermediate filaments to microtubules[17]. Phosphorylation of MAPs, including tau protein,

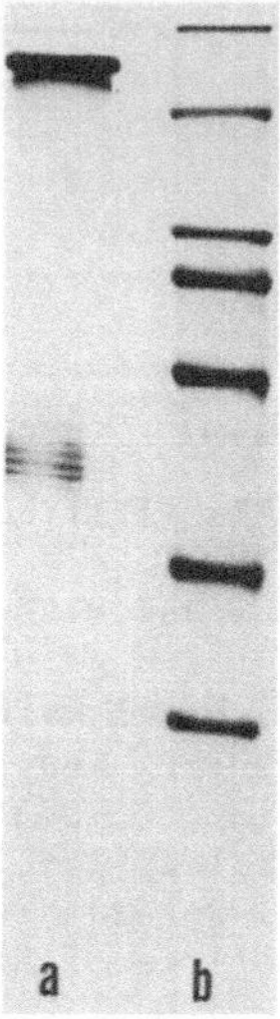

Fig. 1. Coomassie blue stained SDS-polyacrylamide gel electrophoretic profile of bovine heat-stable microtuble preparation. Molecular weight standards at right. (from Yen S-H, et al.: Am J Pathol 126:81, 1987. Used with permission.)

inhibits their ability to promote microtubule assembly[15,16]. In Alzheimer's disease there is preliminary evidence that phosphorylation states of cytoskeletal proteins may be altered[18,19].

In this report we summarize our studies of ANT using two groups of monoclonal antibodies, and critically assess the evidence that suggests that ANT are composed of heat-stable MAPs. We have taken advantage of the hybridoma technique of Kohler and Milstein[20] to produce mouse monoclonal antibodies that are immunoreactive with ANT. The antibodies we have generated and characterized include a group of ten monoclonal antibodies raised against a tangle-enriched preparation. Half of the preparation was fixed in formaldehyde before immunizing the mice[21]. This group of antibodies is referred to as "anti-ANT". Among a second group of monoclonal antibodies raised against a crude homogenate of basal forebrain from Alzheimer brain[22], are six monoclonal antibodies that react with high molecular weight neurofilament proteins and immunostain axons[23]. Individual antibodies of this group are designated by the prefix "NP."

RESULTS AND DISCUSSION

Some Anti-ANT Antibodies Recognize Unique Antigenic Determinants

The ten anti-ANT antibodies were characterized by immunological methods. All antibodies were of the IgG class. They all bound to ANT in preparations of isolated perikarya from Alzheimer brain[21] as judged by double-labeling with thioflavine S, a fluorochrome that binds to ANT[24]. However, they were not equally reactive with ANT. The proportion of ANT labeled varied from 25 to 100 percent. Antibodies that bound to the greatest proportion of ANT, in most cases recognized antigenic determinants that were unique to ANT (Ab39, Ab64, Ab69, Ab117), since they did not bind to any normal brain proteins on Western-type immunoblots. The immunoreactivity of these antibodies to ANT could only be absorbed with preparations that contained ANT and not by normal proteins.

Immunoperoxidase electron microscopic studies with Ab39 and Ab69 showed that their epitopes were localized to paired helical filaments of ANT[25]. Recent immunogold studies have confirmed this with Ab39 (Fig. 2).

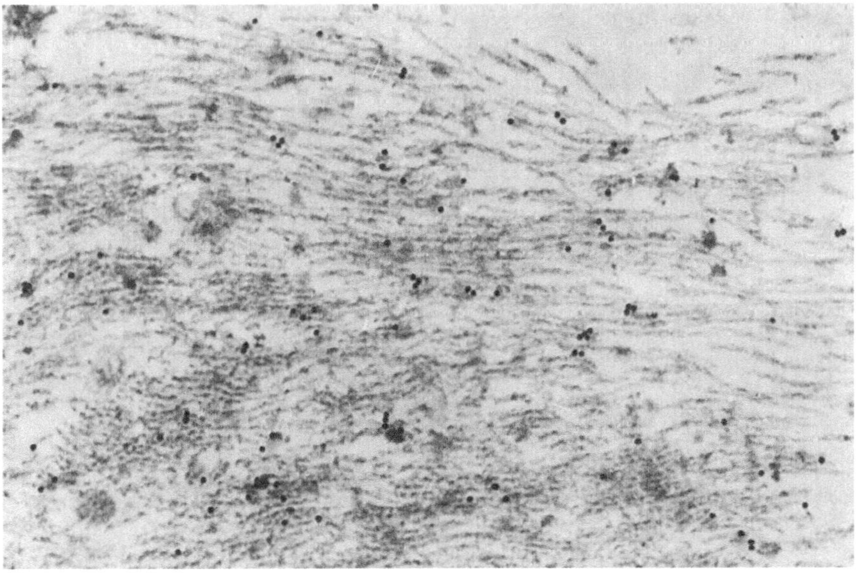

Fig. 2. Immunogold staining of Lowacryl embedded tissue section from a brain biopsy of a patient with Alzheimer's disease. Gold particles immunodecorate PHF. (51,000 X).

Our data suggest that the epitopes recognized by these antibodies are intrinsic to PHF, since these antibodies react with detergent-treated ANT. We are intrigued by the possibility that these unique epitopes may be present on proteins that are not normally expressed in the adult brain; however, we have not excluded the possibility that they may represent PHF-specific, conformational epitopes due to modification of normal neuronal proteins. It will be important to screen human brain cDNA expression vector libraries with these antibodies in an attempt to find a transcript that encodes for theses epitopes[26].

Some Anti-ANT Antibodies Recognize Antigenic Determinants in MAPs

On Western-type immunoblots of Alzheimer brain homogenates, all anti-ANT antibodies bound to proteins at the top of the 4% acrylamide gel[21], which contain PHF. Five of the ten anti-ANT antibodies also bound to soluble proteins in either Alzheimer or normal brain preparations[21]. These soluble proteins were extractable with physiological buffers and formed several bands in the 50-70 kd molecular weight range on SDS polyacrylamide gels. Similar proteins were detected in bovine microtubule preparations[27] (Fig. 3). The proteins were stable to heat treatment and co-localized with proteins that reacted with polyclonal antisera to tau protein. Interestingly, four of the ten anti-ANT antibodies also showed reactivity with high molecular weight, heat-stable MAPs in bovine microtubule preparations[27]. These proteins were MAP_2 based upon their immunoreactivity to antibodies that were specific to MAP_2. The absence of high molecular weight MAPs from human material was likely due to their rapid postmortem degradation.

The significance of shared epitopes in heat-stable MAPs (MAP_2 and tau protein) is not known. It is tempting to speculate that these epitopes may represent conserved recognition sites that mediate macromolecular interactions. The theoretical possibility that monoclonal antibodies might be able to detect such sites has been previously suggested[28]. In the case of MAPs, this is an attractive hypothesis since various MAPs might have conserved domains that mediate interaction with microtubules. Epitope mapping will be necessary to prove that these proteins have shared amino acid sequences that may be capable of interacting with microtubules.

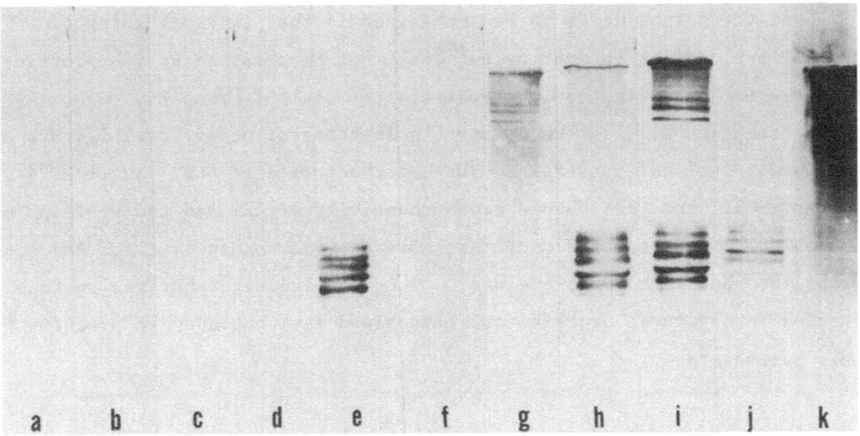

Fig. 3. Western-type immunoblot of bovine, heat-stable microtubule preparation as in Fig. 1 with the ten anti-ANT monoclonal antibodies and an anti-MAP_2 antibody. (a-Ab39, b-Ab64, c-Ab69, d-Ab117, e-Ab175, f-Ab215, g-Ab322, h-Ab635, i-Ab636, j-Ab705, k-anti-MAP_2.) (From Yen S-H, et al.: Am J Pathol 126:81, 1987. Used with permission.)

An important point to be emphasized from these results is that some antibodies generated by immunizing mice with ANT bound to antigenic determinants that were common not only to ANT and tau, but also to MAP_2. Previous studies have shown that a limited number of anti-MAP_2 antibodies bind to ANT[29]. This data is consistent with the hypothesis that MAP_2, or parts of MAP_2, may be incorporated into ANT.

Tau-like Epitopes Are Present in ANT

The fact that five of our ten anti-ANT monoclonal antibodies bind to tau proteins, in addition to evidence from our studies[27] and others[1-5] that anti-tau antibodies bind to ANT (Fig. 4) favors the idea that tau-like epitopes may be the major antigenic component of ANT. Whether ANT are composed of tau proteins or other proteins with tau-like epitopes is currently not resolved. The fact that antisera raised against ANT often contain antibodies that bind to tau proteins[3] is further evidence that tau-like epitopes in ANT are immunodominant. Although it is possible that ANT are composed primarily of an "altered form" of tau protein, the available evidence also supports the possibility that ANT are composed of unique proteins that share epitopes with normal cytoskeletal components, including tau, MAP_2 and neurofilament (see below).

One thing that needs to be emphasized is that immunostaining of ANT by anti-ANT antibodies, in most cases, could not be absorbed by tau proteins or crude heat-stable microtubule preparations[27]. In this respect, our immunoabsorption results are comparable to those of other investigators who have used polyclonal antisera to PHF-enriched preparations, "anti-PHF"[30]. The "anti-PHF" antisera showed cross-reactivity of ANT and tau proteins on immunoblots of SDS-treated proteins; however, immunoreactivity of ANT could not be absorbed with tau proteins[31]. This may indicate that the epitope detected in denatured proteins on immunoblots may be hidden in proteins in their native state.

Some investigators have stated that antibodies raised against preparations from Alzheimer brain produce antibodies reactive to tau proteins, but not to other cytoskeletal proteins[3]. We were thus interested in further characterizing a group of monoclonal antibodies (NP series), six of which reacted with neurofilaments and ANT (ANT-reactive anti-NF antibodies), produced by immunizing mice with Alzheimer brain homogenates[22].

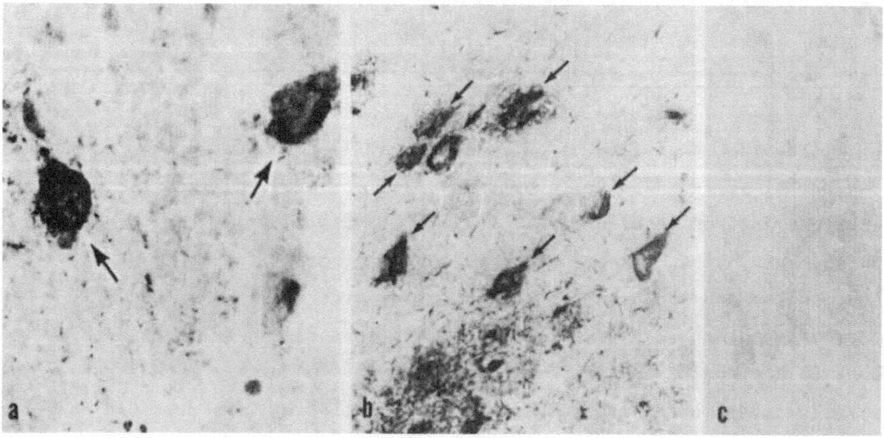

Fig. 4. Immunoperoxidase staining of frozen sections from the hippocampus of Alzheimer's disease. Anti-tau antibodies bind to ANT (arrows), neurites in gray matter and in senile plaques (*). Both polyclonal antisera (a) and monoclonal antibodies (Tau-1) (b) react with ANT. Far fewer ANT are detected with polyclonal anti-tau antisera compared to Tau-1. If tissue sections are incubated for 10 min. in 0.01% trypsin, Tau-1 no longer reacts with ANT (c).

ANT-Reactive Anti-NF Antibodies Have Phosphorylated Epitopes

Several studies of ANT have shown that some anti-neurofilament (anti-NF) antibodies bind to ANT[32-38]. We have characterized six monoclonal anti-NF (NP series) using immunocytochemical and immunochemical methods[39]. All six antibodies immunostained axons in white matter and both isolated ANT and ANT in frozen sections[23]. Immunoblots of cytoskeletal preparations with these antibodies showed that they all reacted with the high molecular weight proteins (210 kD and 150 kD) of the neurofilament triplet, but not with the 68 kD neurofilament protein[39]. Immunoreactivity of the antibodies to electroblotted neurofilament proteins was decreased to different degrees after phosphatase treatment[39]. On tissue sections incubated with phosphatase two of the six anti-NF antibodies bound to axons; while four showed absence of axonal staining under the same conditions[23,39]. The discrepancy between immunocytochemical staining and Western blots may be due to phosphate groups in some of the epitopes that are only accessible to the enzyme when neurofilament protein is in the denatured state.

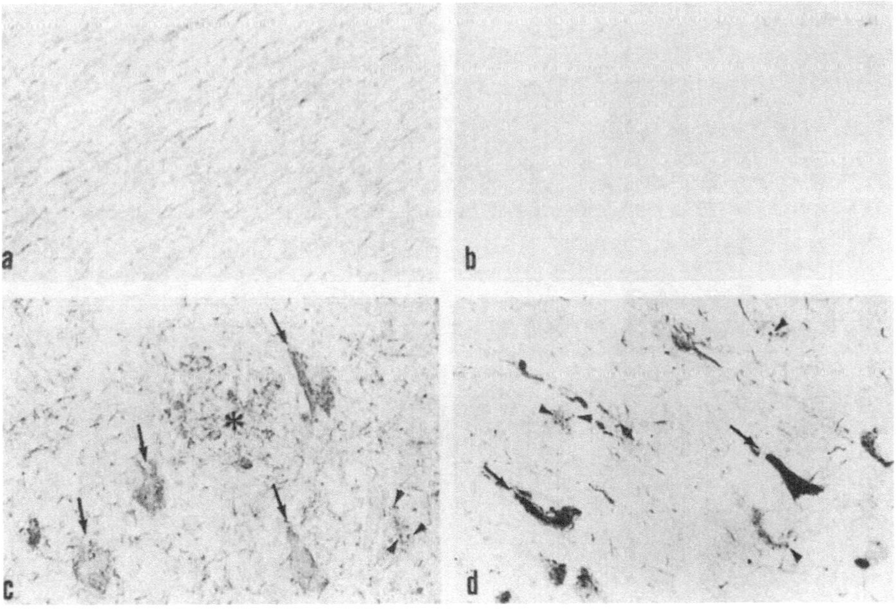

Fig. 5. Immunoperoxidase staining of frozen sections of Alzheimer hippocampus with NP14. Axonal staining in (a) is eliminated by incubating tissue sections with alkaline phosphatase (b), but ANT are stained after either phosphatase (c) or trypsin treatment (d). Trysin enhances ANT staining. In addition to ANT (arrows) and neurites in senile plaques (*), NP14 also immunostains granulovacuolar degeneration (arrow heads).

Phosphatase treatment had no effect on the binding of most of the anti-NF antibodies to ANT[39,40], indicating that the antigenic determinant differed in ANT as compared to normal proteins (Fig. 5). This may indicate that phosphate groups in the epitopes are not accessible to the enzyme active site, possibly due to abnormal folding or packing of the epitope in ANT or that ANT contain some material such as aluminum[41] that may inhibit the enzyme.

ANT-Reactive Anti-NF Antibodies Recognize Tau Protein

Microtuble preparations were used to screen a series of anti-NF monoclonals, including the NP series, two well-characterized antibodies that react with ANT (Anderton's RT97[34] and Sternberger's 07-5[42]) and three anti-NF antibodies that do not react with ANT (Sternberger-Meyer, Inc. (SMI)-32, Behringer-Mannheim (BM) 200 and 160)[39]. Interestingly, all ANT-reactive anti-NF antibodies reacted with tau proteins in bovine microtubule preparations (Fig. 6). Some of the ANT-reactive anti-NF antibodies also recognized human tau proteins[39]. That the proteins were really tau proteins and not degradation products of neurofilament was demonstrated by binding of co-migrating proteins with antibodies that recognize tau proteins (Tau-1 and anti-ANT Ab635).

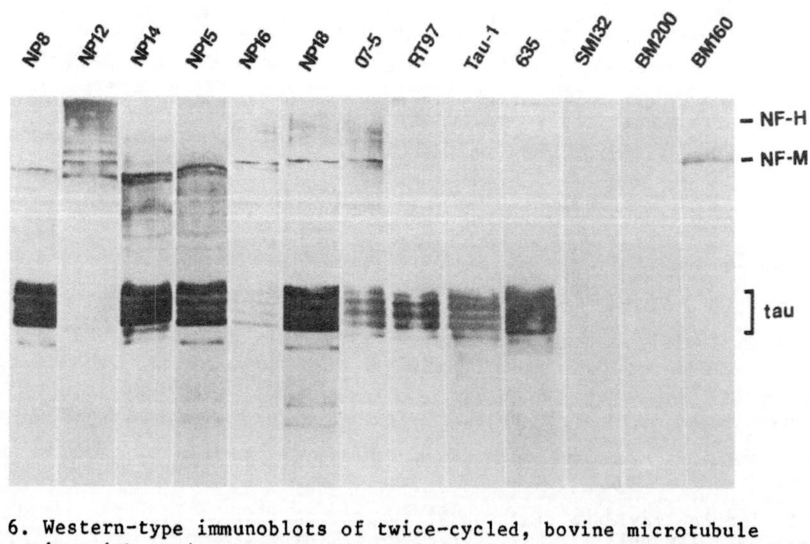

Fig. 6. Western-type immunoblots of twice-cycled, bovine microtubule preparation with anti-NF monoclonal antibodies. Only those antibodies that bind to ANT recognize tau proteins. The microtubule contained a small amount of neurofilament. (From Ksiezak-Reding H., et al.: Proc. Natl. Acad. Sci. (USA) in press(1987).

The significance of cross reaction between tau protein and neurofilment was emphasized further by the observation that anti-NF antibodies that fail to show cross reactivity between neurofilament and tau proteins do not bind to ANT[39]. The role of neurofilament in the formation of ANT, therefore, needs to be re-evaluated.

It may be questioned whether the antibodies of the NP series are indeed anti-NF. Immunocytochemical studies suggest that this is the case, since they immunostain axons in post-mortem human brain sections. On the other hand, well-characterized antibodies to tau protein do not stain axons from human tissue. This may be due to the fact that microtubules are rapidly depolymerized in autopsy material.

Immunoperoxidase staining of ANT in frozen sections from Alzheimer hippocampi by ANT-reactive anti-NF antibodies was enhanced by brief incubation of the sections with trypsin (Fig. 5). In contrast, brief trypsin treatment destroyed the immunoreactivity of ANT with the anti-tau antibody Tau-1 (Fig. 4). Thus, nonphosphorylated tau protein epitopes[14] are labile to trypsin treatment, while other epitopes, in particular the

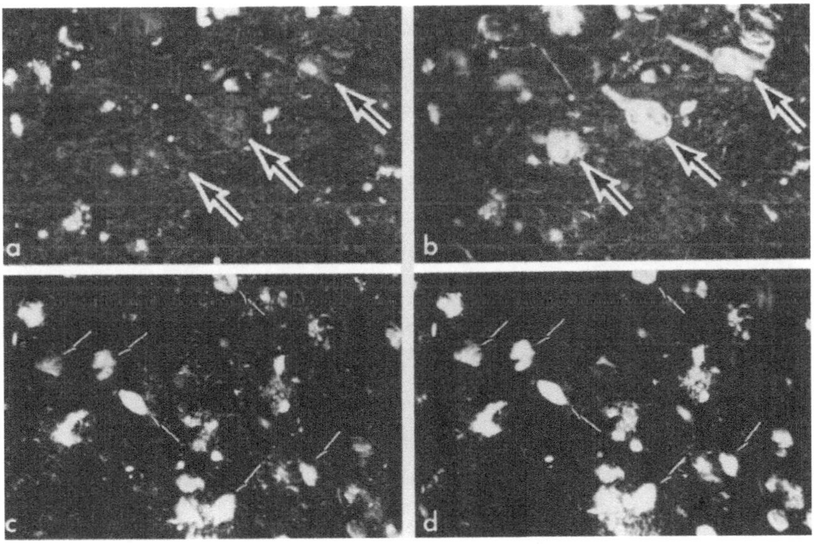

Fig. 7. Double labeling immunofluorescence of paraffin-embedded hippocampus stained with Tau-1 and counterstained with thioflavine S. Note absence of Tau-1 reactivity (a) with ANT (arrows) labeled by thioflavine S (b). On frozen sections Tau-1 binds to ANT (arrows) in (c) similar to thioflavine S (d).

phosphate-dependent epitopes common to neurofilament and tau proteins (and epitopes recognized by anti-ANT), are resistant to trypsin. Previous studies have shown that protease treatment does not alter the morphology of PHF [43]. This raises the question of whether tau proteins are essential in the maintenance of PHF. The detection of tau-like epitopes in SDS-treated ANT [27], nevertheless, argues that tau-like epitopes are integral components of PHF.

The inability of a specific anti-tau monoclonal antibody (Tau-1)[14] to bind to ANT in paraffin sections (Fig. 7) unless they were pretreated with phosphatase has been the main argument that tau protein is abnormally phosphorylated in Alzheimer's disease[19]. Further analytical studies that measure the actual phosphate content of tau proteins in Alzheimer's disease, similar to that which has been done for neurofilament[44], will be needed to prove this point. Since we have been able to demonstrate Tau-1 immunostaining of ANT in frozen sections without phosphatase treatment (Fig. 4 and 7), one wonders whether the necessity to use phosphatase treatment for immunostaining paraffin sections is a consequence of tissue processing.

CONCLUDING REMARKS

Our immunocytochemical studies show that antibodies that recognize antigenic determinants in normal cytoskeletal proteins bind to only some ANT[32] (Fig. 8). In contrast, antibodies that react with tangle-specific epitopes bind to the majority of ANT (Fig. 8). These results point out an important property of ANT that has not been emphasized, their heterogeneity.

Alzheimer neurofibrillary tangles are heterogeneous by morphological, biochemical, and immunological criteria. At the light microscopic level, ANT are present in both small non-pyramidal neurons, where they are little more than coarse perinuclear fibrils, and in large pyramidal neurons, where they form compact, flame-shaped or globose aggregates. With the hematoxylin and eosin stain most ANT are amphophilic to basophilic, but some are eosinophilic. Some ANT are stained well with thioflavine and silver methods, while others are stained much weaker.

Ultrastructural studies of ANT confirm their heterogeneity. Some ANT are composed of loosely arranged filaments and others are dense, almost paracrystalline arrays. Although the majority of the filaments are

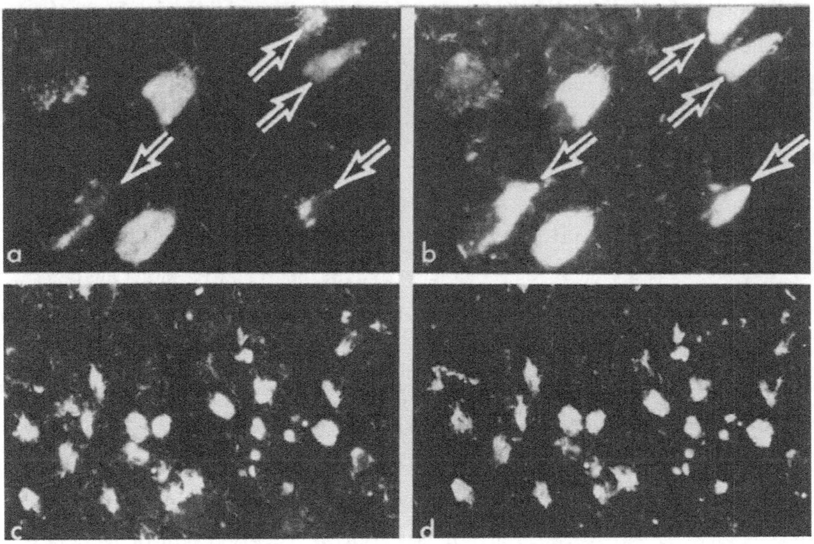

Fig. 8 Double staining immunofluorescence of frozen sections using NP18 (a) and Ab39 (c) and thioflavine S staining of same sections in (b) and (d). Anti-ANT bind to virtually every ANT while some ANT are not reactive with anti-NF (arrows).

PHF, other filaments including 15-18 nm diameter straight filaments[45], PHF with a longer periodicity[46] and normal appearing intermediate filaments are also present to varying degrees in different ANT. In addition ANT almost always contains electron-dense granular material. Very few studies have addressed the nature of the nonfilamentous components of ANT.

Biochemical studies have shown that some ANT are composed of insoluble material that resists harsh acid or alkaline treatment or enzymatic digestion[43,47], but others contain soluble and extractable elements. The former material fails to enter the stacking gel of acrylamide gel electrophoresis[47], while the latter material either enters the gel forming a diffuse smear or a set of discrete protein bands in the mid-molecular weight region[48]. Characterizing the biochemical composition of ANT must bear in mind this heterogeneity. ANT purified by methods that take advantage of the insolubility characteristics may be lacking in extractable elements, while methods that yield ANT enriched in soluble components, may produce quite conflicting results.

Because of these inherent properties of ANT, biochemical studies of ANT may be intractable. Immunochemical methods may prove to be the most reliable method of "dissecting" the ANT. It is important in this respect, however, that antibodies be well-characterized and their limitations kept in mind.

What then can be said for the molecular nature of ANT? The definitive answer awaits studies that address the evolution of ANT. What must be resolved is whether the heterogeneity that is observed with morphological, biochemical and immunological methods is due to developmental ("maturational") properties of ANT or to factors such as the size or type of neuron affected. A study of neurofibrillary degeneration in Down's syndrome patients of various ages with the various antibodies to which ANT have shown at least some reactivity may be able to detect early neuronal changes that are precursors to ANT formation. Of particular interest would be the abnormal disposition of heat-stable MAPs in vulnerable neurons of Down's syndrome patients before they would be expected to have Alzheimer type changes[49]. This might suggest that MAPs are involved in the pathogenesis of ANT.

A better understanding of ANT would come from an elucidation of the nature of the unique epitopes recognized by anti-ANT monoclonal antibodies. Do these represent unique PHF proteins or unique combinations or alterations of normal cytoskeletal proteins? It may be significant that ANT contain epitopes shared with neurofilament and MAPs. Are these epitopes conserved sequences that are common to several proteins, including possible unique PHF proteins? Finally, it is of interest that many of the epitopes shared between ANT and normal proteins are phosphorylated. Could this indicate that ANT result from alterations in activities of specific kinases?

Acknowledgments: The authors express their thanks to Drs. Binder, Anderton, Sternberger, Shelanski and Davies for the use of their antibodies in this study. This work is supported by NIH grants AG1136, AG4145 and NS7098.

REFERENCES

1. Brion JP, Van den Bosch de Aguilar P, Flament-Durand J: Senile dementia of the Alzheimer type: Morphological and immunocytochemical studies in senile dementia of Alzheimer type, in Traber J, Gispen WH (eds): Advances in Applied Neurological Science, vol 2. Berlin, Springer, 1985, pp 164-174.

2. Grundke-Iqbal I, Iqbal K, Tung Y-C, et al.: Microtubule-associated protein tau. A component of Alzheimer paired helical filaments. J Biol Chem 261:6084-6089, 1986.

3. Kosik KS, Joachim CL, Selkoe DJ: Microtubule-associated protein tau is a major antigenic component of paired helical filaments in Alzheimer disease. Proc Natl Acad Sci (USA) 83:4044-4048, 1986.

4. Wood JG, Mirra SS, Pollock NJ, et al.: Neurofibrillary tangles of Alzheimer disease share antigenic determinants with axonal microtubule-associated protein tau. Proc Natl Acad Sci (USA) 83:4040-4043, 1986.

5. Iqbal K, Zaidi T, Wen GY, et al.: Defective brain microtubule assembly in Alzheimer's disease. Lancet ii:421-426, 1986.

6. Weingarten MD, Lockwood AH, Kirschner MW: A protein factor essential for microtubule assembly. Proc Natl Acad Sci (USA) 72:1858-1862, 1975.

7. Kidd M: Paired helical filaments in electron microscopy in Alzheimer's disease. Nature (London) 197:192-194, 1963.

8. Cleveland DW, Hwo S-Y, Kirschner MW: Physical and chemical properties of purified tau factor and the role of tau in microtubule assembly. J Mol Biol 116:227-247, 1977.

9. Fellous A, Ohayon R, Mazie J-C, et al.: Tau microheterogeneity: An immunological approach with monoclonal antibodies. (In Soifer D (ed): Dynamic Aspects of Microtubule Biology) Ann New York Acad Sci 446:240-256, 1986.

10. Drubin DC, Caput D, Kirschner MW: Studies on the expression of the microtubule-associated protein, tau, during mouse brain development with newly isolated complementary DNA probes. J Cell Biol 98:1090-1097, 1984.

11. Lindwall G, Cole RD: The purification of tau protein and the occurrence of two phosphorylation states of tau in brain. J Biol Chem 259:12241-12245, 1984.

12. Butler, M, Shelanski, M: Microheterogeneity of microtubule-associated tau proteins is due to differences in phosphorylation. J Neurochem 47:1517-1523, 1986.

13. Drubin D, Kobayashi S, Kirschner M: Association of tau protein with micortubule in living cells. (In Soifer D (ed): Dynamic Aspects of Microtubule Biology) Ann New York Acad Sci 446:257-268, 1986.

14. Binder LI, Frankfurter A, Rebhun LI: The distribution of tau in the mammalian central nervous system. J Cell Biol 101:1371-1378, 1985.

15. Lindwall G, Cole RD: Phosphorylation affects the ability of tau protein to promote microtubule assembly. J Biol Chem 259:5301-5305, 1984.

16. Vallee RB: MAP 2 (Microtubule-associated protein 2), in Shay JW (ed): Cell and Muscle Motility, vol 5. New York, Plenum Press, 1984, pp 289-311.

17. Bloom GS, Vallee RB: Association of microtubule-associated protein 2 (MAP 2) with microtubules and intermediate filaments in cultured brain cells. J Cell Biol 96:1523-1531, 1983.

18. Sternberger NH, Sternberger LA, Ulrich J: Aberrant neurofilament phosphorylation in Alzheimer's disease. Proc Natl Acad Sci USA 82:4274-4276, 1985.

19. Grundke-Iqbal I, Iqbal K, Tung Y-C, et al.: Abnormal phosphorylation of the microtubule-associated protein t (tau) in Alzheimer cytoskeletal pathology. Proc Natl Acad Sci (USA) 83:4913-4917, 1986.

20. Kohler G, Milstein C: Continuous cultures of fused cells secreting antibodies of predefined specificity. Nature (London) 256:495-497, 1975.

21. Yen S-H, Crowe A, Dickson DW: Monoclonal antibodies to Alzheimer neurofibrillary tangles 1. Identification of polypeptides. Am J Pathol 120:282-291, 1985.

22. Wolozin BL, Pruchnicki A, Dickson DW, et al.: A neuronal antigen in the Alzheimer brain. Science 232:648-650, 1986.

23. Dickson DW, Yen S-H, Davies P: Alzheimer neurofibrillary tangles contain phosphorylated and nonphosphorylated neurofilament epitopes recognized by monoclonal antibodies (abstract). Am J Clin Pathol 85:381, 1986.

24. Schwartz P: Amyloidosis--Cause and Manifestation of Old Age. Springfield, Illinois, Thomas, 1971.

25. Dickson DW, Kress Y, Crowe A, et al.: Monoclonal antibodies to Alzheimer neurofibrillary tangles (ANT) 2. Demonstration of a common antigenic determinant between ANT and neurofibrillary degeneration in progressive supranuclear palsy. Am J Pathol 120:292-302, 1985.

26. Dammerman M, Yen S-H, Shafit-Zagardo B: A human cDNA encoding epitopes for MAP_2 and Alzheimer neurofibrillary tangles (abstract). American Society for Cell Biology Summer Research Conference, Molecular Biology of the Cytoskeleton, ST22, 1986.

27. Yen S-H, Dickson DW, Crowe A, et al.: Alzheimer's neurofibrillary tangles contain unique epitopes and epitopes in common with the heat-stable microtubule associated proteins tau and MAP_2. Am J Pathol 126:81-91, 1987.

28. Lane D, Koprowski H: Molecular recognition and the future of monoclonal antibodies. Nature (London) 296:200-202, 1982.

29. Kosik KS, Duffy LK, Dowling MM, et al.: Microtubule-associated protein 2: Monoclonal antibodies demonstrate the selective incorporation of certain epitopes into Alzheimer neurofibrillary tangles. Proc Natl Acad Sci (USA) 81:7941-7945, 1984.

30. Ihara Y, Abraham C, Selkoe DJ: Antibodies to paired helical filaments in Alheimer's disease do not recognize normal brain proteins. Nature (London) 304:727-730, 1983.

31. Nukina N, Ihara Y: One of the antigenic determinants of paired helical filaments is related to tau protein. J Biochem (Tokyo) 99:1541-1544, 1986.

32. Yen S-H, Gaskin F, Terry RD: Immunocytochemical studies of neurofibrillary tangles. Am J Pathol 104:77-89, 1981.

33. Anderton BH, Brienburg D, Downes MJ, et al.: Monoclonal antibodies show that neurofibrillary tangles and neurofilaments share antigenic determinants. Nature (London) 298:84-86, 1982.

34. Autillo-Gambetti L, Gambetti P, Crane RC: Paired helical filaments: Relatedness to neurofilament shown by silver staining and reactivity with monoclonal antibodies, in Katzman R (ed): Biological Aspects of Alzheimer's Disease, Banbury Report, vol 15. Cold Spring Harbor, New York, Cold Spring Harbor Laboratory, 1983, pp 117-124.

35. Elovaara I, Paetau A, Lehto V-P, et al.: Immunocytochemical studies of Alzheimer neuronal perikarya with intermediate filament antisera. J Neurol Sci 62:315-326, 1983.

36. Perry G, Rizzuto N, Autilio-Gambetti L, et al.: Paired helical filaments from Alzheimer disease patients contain cytoskeletal components. Proc Natl Acad Sci (USA) 82:3916-3920, 1985.

37. Miller CCJ, Brion J-B, Calvert R, et al.: Alzheimer's paired helical filaments share epitopes with neurofilament side arms. EMBO J 5:269-276, 1986.

38. Ksiezak-Reding H, Yen S-H: Two monoclonal antibodies recognize Alzheimer's neurofibrillary tangles, neurofilament and microtubule associated proteins. J Neurochem 48:455-462, 1987.

39. Ksiezak-Redding H, Dickson DW, Yen S-H: Recognition of tau epitopes by anti-neurofilament antibodies that stain Alzheimer neurofibrillary tangles. Proc Natl Acad Sci (USA) (in press), 1987.

40. Yen S-H, Dickson DW, Peterson C, et al.: Cytoskeletal abnormalities in neuropathology, in Zimmerman HM (ed): Progresss in Neuropathology, vol 6. New York, Raven Press, 1986, pp 63-90.

41. Perl DP, Munoz-Garcia D, Good PF, et al.: Calculation of intracellular aluminum concentration in neurofibrillary tangle (NFT) bearing and NFT-free hippocampal neurons of ALS/Parkinsonism-dementia (PD) of Guam using laser microprobe mass analysis (LAMMA) (abstract). J Neuropathol Exp Neurol 45:379, 1986.

42. Sternberger LA, Sternberger NH: Monoclonal antibodies distinguish phosphorylated and nonphosphorylated forms of neurofilament in situ. Proc Natl Acd Sci (USA) 80:6126-6130, 1983.

43. Yen S-H, Kress Y: The effect of chemical reagents or proteases on the ultrastructure of paired helical filaments, in Katzman R (ed): Biological Aspects of Alzheimer's Disease, Banbury Report, vol 15. Cold Spring Harbor, New York, Cold Spring Harbor Laboratory, 1983, pp 155-165.

44. Ksiezak-Redding H, Yen S-H: Phosphatase and carbocyanine dye-binding defines different types of phosphate groups in mammalian neurofilaments (abstract). J Cell Biol 103:421a, 1986.

45. Yagishita S, Itoh Y, Nan W, et al.: Reappraisal of the fine structure of Alzheimer's neurofibrillary tangles. Acta Neuropathol (Berl) 54:239-246, 1981.

46. Brion JP, Couck AM, Flament-Durand J: Ultrastructural study of enriched fractions of "tangles" from human patients with senile dementia of the Alzheimer type. Acta Neuropathol (Berl) 64:148-152, 1984.

47. Selkoe DJ, Ihara Y, Salazar FJ: Alzheimer's disease: Insolubility of partially purified paired helical filaments in sodium dodecyl sulfate and urea. Science 215:1243-1245, 1982.

48. Iqbal K, Zaidi T, Thompson CH, et al.: Alzheimer paired helical filaments: Bulk isolation, solubility, and protein composition. Acta Neuropathol (Berl) 62:167-177, 1984.

49. Coyle JT, Oster-Granitz ML, Gearhart JD: The neurobiologic consequences of Down syndrome. Brain Res Bull 16:773-787, 1986.

AN IMMUNOCYTOCHEMICAL COMPARATIVE ANALYSIS OF TAU IN NEURODEGENERATIVE DISEASES

Nancy J. Pollock, Suzanne S. Mirra and John G. Wood

Depts. of Anatomy and Cell Biology and
Pathology and Laboratory Medicine and the V.A.
Emory University School of Medicine, Atlanta, GA 30322

INTRODUCTION

Microtubules are essential for maintaining neuronal function. As a structural component, they serve to strengthen the cytoplasmic matrix by interacting with other cytoskeletal elements through microtubule-associated proteins (MAPs) (1-4). The formation and maintainance of neuritic processes in vitro is dependent upon the presence of polymerized microtubules (5). By providing an avenue for the vesicular translocator, kinesin, (6) microtubules are also required for fast axonal transport. The diversity of functions performed by microtubules is a reflection of their heterogeneity. They are composed of alpha and beta tubulin subunits (7), which exhibit increasing heterogeneity during development (8), leading to at least 11 tubulin isoforms (9). Posttranslational modifications of tubulin such as acetylation (10), phosphorylation (11,12), tyrosinolation (13) and detyrosinolation (14) also serve to diversify microtubules. The association of various MAPs with neuronal microtubules adds to the complexity of this population of cytoskeletal elements. Although microtubules are heterogeneous, they are distributed throughout the neuron. The interaction with MAPs, which are apparently not equally parceled among neuronal compartments, provides a mechanism for stabilizing and maintaining subpopulations of microtubules required for specific functions. The segregation of MAPs, or different forms of individual MAPs, within specific neuronal compartments implies the neuron possesses mechanisms for keeping them apart. Alteration of the sorting and segregation process necessary to maintain neuronal compartmentalization, could result in loss of neuronal form as well as function. Alzheimer's disease-affected neurons are one model for the study of improper neuronal organization of cytoskeletal elements.

COMPARTMENTALIZATION OF ALZHEIMER'S DISEASE-AFFECTED NEURONS

Alzheimer's disease (AD) is a neurodegenerative disease in which two major pathological features are present: the neurofibrillary tangle (NFT) and the neuritic plaque (NP). The NFT is found in the perikaryon, whereas the NP is a spherical structure composed of neuritic processes at its periphery, often surrounding an amyloid core. Electron microscopic studies determined that NFT and many plaque neurites are composed of paired helical filaments (PHF) (15,16). PHF have been shown to share antigenic determinants with neurofilament polypeptides (17-20), and other cytoskeletal components (21-25), yet still may contain epitopes not found in normal brain (26). Recent studies however, show a constituent of the NFT and NP to be a modified form of the microtubule-associated protein, tau (27,28,29). The localization of tau to the NFT in the perikaryon suggests the atypical segregation of this protein in disease-affected neurons, since tau has been reported to be concentrated in axons (30).

Alzheimer's disease is only one of many neurodegenerative diseases characterized by filamentous cytoplasmic aggregates. Other neurodegenerative diseases which exhibit such filamentous structures include: Pick's disease, progressive supranuclear palsy (PSP) and amyotrophic lateral sclerosis-parkinson dementia complex of Guam (ALS-PDC of Guam). Parkinson's disease (PD), a movement disorder which may be associated with dementia, also exhibits a characteristic cytoplasmic inclusion, the Lewy body. We were interested in investigating the association of tau with other cytoplasmic inclusions. Therefore, using immunocytochemistry, we compared these cytoplasmic structures using tissue from a spectrum of neurodegenerative diseases for alterations of the neuronal cytoskeleton.

TAU IMMUNOLOCALIZATION IN PICK BODIES AND GLOBOSE NEUROFIBRILLARY TANGLES OF PSP

Immunocytochemistry was performed on formalin fixed paraffin tissue sections using two monoclonal antibodies, Tau-1 and Tau-2 (kindly provided by Dr. L.I. Binder), which are directed against bovine tau. Tau-2 immunocytochemistry was performed at 1:100 and Tau-1 was used at 1:500, dilutions optimal for immunoreactivity of pathological structures; axonal staining however, was not appreciated. The lack of axonal label may be due to decreased antigenicity of tau after formalin fixation. In contrast, the antigenicity of modified tau associated with pathological structures is apparently preserved, perhaps by the stability of the microenviornment. Although Tau-2 recognizes a different epitope on the tau molecule (31) than the previously described Tau-1 antibody (30), it does recognize human tau. Tau-2 therefore, was used for the staining of human tissue sections and compared with routine histological silver impregnation and hematoxylin and eosin (H&E) stains, as described below.

Pick bodies, characteristic argentophilic cytoplasmic inclusions, found in neurons in Pick's disease, contain straight filaments averaging 15nm in diameter (32). Most globose neurofibrillary tangles, which characterize PSP disease tissue, also contain 15nm straight filaments (33), although some globose NFTs contain paired helical filaments (PHF) (34), similiar to those found in Alzheimer's NFTs (15,16) or a combination of both straight filaments and PHF. PSP globose NFT share antigenic determinants with Alzheimer NFTs (35). Pick bodies have epitopes in common with both neurofilaments (36) and PHF (37). We examined paraffin sections of tissue from cases of Pick's disease and PSP for tau immunoreactivity. The Pick bodies present in the pyramidal cells and dentate neurons of the hippocampus and the globose NFTs in the subthalamic and inferior olivary nuclei all exhibited Tau-2 immunoreactivity. Tau-positive Pick bodies (Fig.1) and globose neurofibrillary tangles (Fig.2) can be compared to corresponding silver stained sections (Figs. 1 and 2).

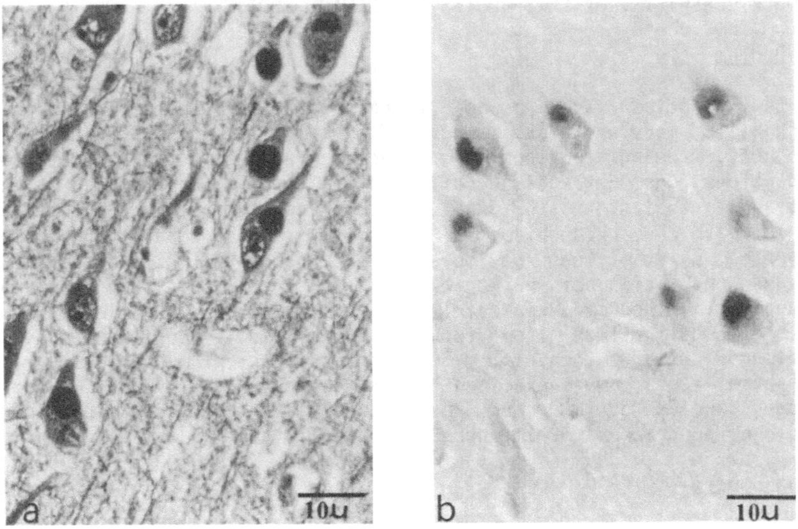

Figure 1.

Light micrographs of paraffin sections from formalin-fixed Pick's disease tissue. A) Silver stained section showing argentophilic Pick bodies in the pyramidal cells of the hippocampus. B) Tau-2 immunopositive Pick bodies in hippocampal pyramidal cells from an adjacent section.

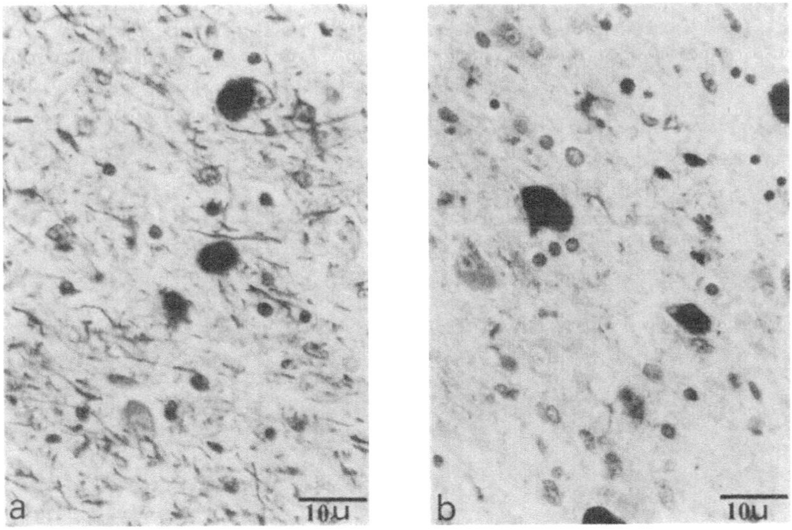

Figure 2.

Light micrographs of paraffin sections from PSP tissue illustrating globose neurofibrillary tangles. A) Subthalamic nucleus neurons from PSP tissue bearing globose neurofibrillary tangles seen by routine silver stain. B) Tau-2 immunopositive globose neurofibrillary tangles in neurons in the subthalamic nucleus.

Lewy Bodies

Lewy bodies are round intracytoplasmic inclusions with an eosinophilic core and a paler surrounding halo. Immuno-light and electron microscopy, using antibodies directed against the neurofilament polypeptides, demonstrate neurofilament epitopes as a component of these cytoplasmic inclusions (38,39). Although Lewy bodies are characteristic of idiopathic parkinsonism, a progressive neurodegenerative disease manifested by the loss of dopaminergic neurons in the nigrostriatal pathway, they are not pathognomonic. In fact, they are frequently encountered in neuromelanin-bearing neurons in Alzheimer's disease brains (40). Tissue slices from such a histologically-confirmed case of Alzheimer's disease, containing Lewy bodies in the pigmented cells of the substantia nigra, were examined for tau immunoreactivity. Unlike the the globose tangles of PSP and the Pick bodies, cytoplasmic Lewy bodies contain no apparent Tau-2 immunoreactivity (Fig. 3).

Hirano Bodies

Hirano bodies, first described in the ALS-parkinson-dementia complex of Guam (41), are rod shaped, eosinophilic structures. Ultrastructurally, these cytoplasmic inclusions are paracrystalline arrays of densely packed filaments located within neuritic processes and occasionally within neuronal perikarya. Immunocytochemical studies have implicated actin as their major constituent (42), however Hirano bodies also share antigenic determinants with α-actinin and tropomyosin (43). In addition to being found in ALS-PD complex of Guam, they have also been described in normal aged people (44,45), in Alzheimer's disease (45) and Pick's disease (46,47). The coexistence of Hirano bodies in the hippocampus of many disorders with tau-positive cytoplasmic inclusions, led us to investigate their tau immunoreactivity. Hirano bodies in the hippocampus in a Pick's disease case, did not show any apparent Tau-2 positive immunoreactivity (Fig. 4).

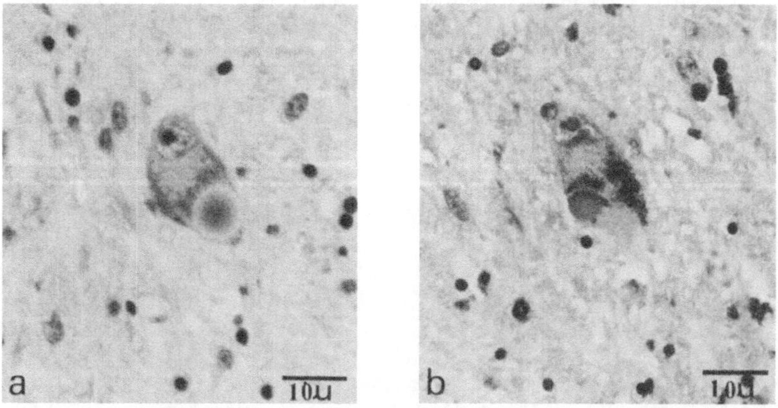

Figure 3.

Light micrographs of paraffin sections from an Alzheimer's disease case showing Lewy bodies in the pigmented cells of the substantia nigra. A) Hematoxylin and eosin stain illustrates the dense core and paler surrounding halo of the Lewy body. B) No labelling of the Lewy bodies is seen after Tau-2 immunoperoxidase staining of the tissue section. The neuromelanin pigment contained within the neuronal cytoplasm should not be confused with immunoperoxidase reaction product.

TAU IMMUNOCYTOCHEMISTRY OF ALZHEIMER NFT IN ALS-PARKINSON-DEMENTIA COMPLEX OF GUAM

The ALS-PDC of Guam is characterized by neurofibrillary tangles of the Alzheimer type, found in widespread neurons including those of the hippocampus, midbrain, brainstem and cerebellum. These NFT are identical in light and electron microscopic appearance to those of Alzheimer's disease; neuritic plaques however, are rarely found in ALS-PD complex (48). Since NFT of ALS-parkinson-dementia complex of Guam tissue are composed of PHF (41), and Alzheimer NFT and NP in aldehyde fixed tissue contain a modified form of tau (27,28,29), we examined paraffin-embedded hippocampus from two cases of ALS-parkinson-dementia complex of Guam for tau immunoreactivity. In the first case, Tau staining was not apparent (data not shown). The lack of staining in one case may reflect the age of the specimen. In the second case, in general, fewer NFT were tau positive than stained with silver (Fig. 5).

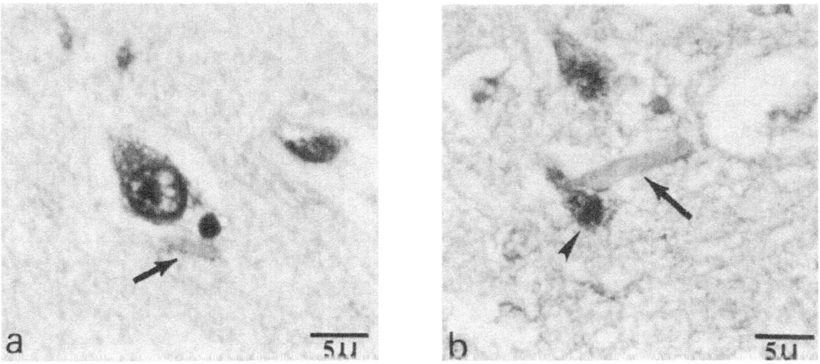

Figure 4.

Light micrographs of paraffin sections illustrating Hirano bodies in the pyramidal cell layer of the hippocampus from Pick's disease tissue. A) The arrow points to a Hirano body in the neuropil as seen by hematoxylin and eosin stain. B). An immunostained section showing a Tau-2 negative Hirano body (arrow) adjacent to a pyramidal neuron containing a Tau-2 immunopositive Pick body (arrowhead).

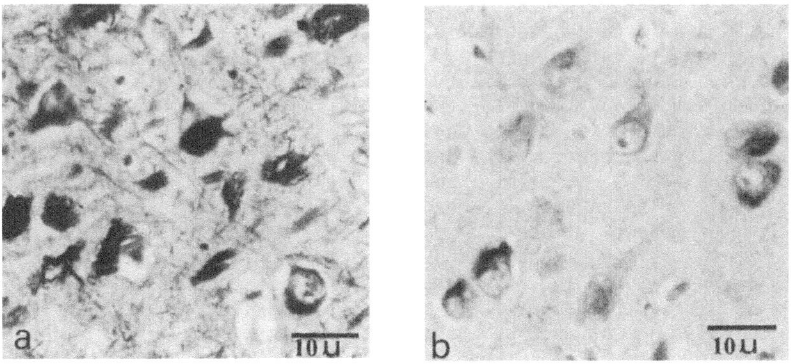

Figure 5.

Light micrographs of paraffin sections from a case of ALS-PD complex. A) Many neurofibrillary tangles of the Alzheimer type in the pyramidal cells of the hippocampus are seen by silver stain. B) Fewer tangle-bearing neurons are tau positive than stain with silver.

DISCUSSION

In this study, we performed a comparative immunocytochemical analysis of tissue from a spectrum of neurodegenerative diseases to determine whether the pathological filamentous inclusions characteristic of each, share antigenic determinants with the microtubule-associated protein, tau. We investigated Pick's disease characterized by Pick bodies, progressive supranuclear palsy (PSP) with globose neurofibrillary tangles, an Alzheimer's case with prominent Lewy bodies (inclusions strongly associated with idiopathic Parkinson's disease) and amyotrophic lateral sclerosis-parkinson dementia complex of Guam cases containing neurofibrillary tangles of the Alzheimer type. Hirano bodies, cytoplasmic inclusions often found in the hippocampus in most of the diseases described above, were also examined. Our results show that Pick bodies as well as NFT of Alzheimer's disease, ALS-PD complex of Guam and PSP, share antigenic determinants with tau, whereas Lewy bodies and Hirano bodies do not. Consistent with this result, using a different antibody directed against tau, the filamentous aggregates of Pick's disease and PSP were found to share antigenic determinants with a modified form of tau (49) as seen in Alzheimer's disease-affected neurons (27,28,29).

Hirano bodies are intraneuronal structures found in the pyramidal cells of the hippocampus in AD tissue, often in juxtaposition to other pathological structures, e.g. NFT. Anti-tau immunolabelled Hirano bodies therefore, could be confused with a partially immunoreactive NFT in the pyramidal cell perikarya in immunocytochemically prepared tissue. It is for this reason that we chose Hirano body-containing Pick's disease tissue to examine whether Hirano bodies are tau-immunoreactive. Although Pick bodies are also located in hippocampal pyramidal cells, these tau immunopositive spherically shaped structures are less likely to be confused with rod-like structures (44), than a partially immunopositive flamed-shaped neurofibrillary tangle.

The localization of tau to Alzheimer NFT (27,28,29) is unique since several previous studies immunocytochemically demonstrated that Alzheimer NFT contain neurofilament epitopes (17-20). It should be noted, however, that some of the anti-neurofilament antibodies, directed against the phosphorylated epitope of the 200 kd neurofilament polypeptide, used in these studies cross react with other phosphorylated proteins. Regardless of differences in the distribution of cytoskeletal proteins in NFT, the localization of tau to the pathological structures of several neurodegenerative diseases suggests that the aberrant accumulation of tau in disease-affected neurons is a fundamental abnormality. On the other hand, this is not a general phenomenon of neurodegenerative diseases since the Lewy body, characteristic of Parkinson's disease, and Hirano bodies do not appear to exhibit tau-positive immunoreactivity.

The accumulation of tau in the perikarya of disease-affected neurons suggests the aberrant segregation of this microtubule-associated protein, since tau, or a form of tau, has been reported to be concentrated in axons (30). The atypical segregation of tau in disease-affected neurons implies a defect either in the recognition of tau by the transport mechanism, or the transport system itself. Although our data implies a potential defect in the sorting and segregation of tau in disease-affected neurons, it is not able to distinguish between these two possibilities. In either case, since tau acts as a microtubule stabilizing and assembly promoting factor in vitro (50), it is possible that neuronal degeneration results from the instability of axonal microtubules due to the accumulation of tau in the perikarya of diseased neurons. Understanding the mechanisms underlying alterations in the neuronal cytoskeleton during aging and dementia may provide insights into the pathogenesis of several neurodegenerative diseases.

Addendum

Galloway et.al. (51) have recently reported that a population of Hirano bodies in Alzheimer's disease tissue contain tau proteins. In the present results, Hirano bodies in tissue sections from several diseases examined did not show tau immunoreactivity. These immunocytochemical data using several different antibodies directed against tau, indicate that further work on the structural compostion of Hirano bodies is needed.

Acknowledgements

The authors thank Dr. Robert Terry for the Pick's disease cases, Dr. Asao Hirano for the ALS-PDC of Guam material, and Dr. Lester Binder for the monoclonal antibodies.

REFERENCES

1. Pollard TD, Selden SC, Maupin P: Interaction of actin filaments with microtubules. J Cell Biol 99(1):33s-37s, 1984
2. Selden SC, Pollard TD: Interaction of actin filaments with microtubules is mediated by microtubule-associated proteins and regulated by phosphorylation. Ann NY Acad Sci 466:803-812, 1986
3. Williams RC: Interactions of microtubules with neuronal intermediate filaments in vitro. Ann NY Acad Sci 466:798-802, 1986
4. Vallee RB, Bloom GS, Theurkauf WE: Microtubule-associated proteins: Subunits of the cytomatrix. J Cell Biol 99(1):38s-44s, 1984
5. Daniels MP: Colchicine inhibition of nerve fiber formation in vitro. J Cell Biol 53:164-176, 1972
6. Vale RD, Reese TS, Sheetz MP: Identification of a novel force-generating protein, kinesin, involved in microtubule-based motility. Cell 42:39-50, 1985
7. Bryan J, Wilson L: Are cytoplasmic microtubules heteropolymers? Proc Nat'l Acad Sci USA 8(8):1762-1766, 1971
8. Gozes I, Littauer UZ: Tubulin microheterogeneity increases with rat brain maturation. Nature 276:411-413, 1978
9. Denoulet P, Edde B, Jeantet C, et al: Evolution of tubulin heterogeneity during mouse brain development. Biochimie 64:165-172, 1982
10. L'Hernault SW, Rosenbaum JL: Chlamydomonas α-tubulin is posttranslationally modified in the flagella during flagellar assembly. J Cell Biol 97:258-263, 1983
11. Eipper BA: Rat brain microtubule protein: Purification and determination of covalently bound phosphate and carbohydrate. Proc Nat'l Acad Sci USA 69(8):2283-2287, 1972
12. Forgue ST, Dahl JL: Rat brain tubulin: Subunit heterogeneity and phosphorylation. J Neurochem 32:1015-1025, 1979
13. Arce CA, Rodriguez JA, Barra HS, et al: Incorporation of L-tyrosine, L-phenylalanine and L-3,4-dihydoxyphenylalanine as single units into rat brain tubulin. Eur J Biochem 59:145-149, 1975
14. Kumar N, Flavin M: Preferential action of a brain detyrosinolating carboxypeptidase on polymerized tubulin. J Biol Chem 256(14):7678-7686, 1981
15. Kidd M: Paired helical filaments in electron microscopy of Alzheimer's disease. Nature 197:192-193, 1963
16. Terry RD: The fine structure of neurofibrillary tangles in Alzheimer's disease. J Neuropathol Exp Neurol 22:629-642, 1963
17. Dahl D, Selkoe DJ, Pero RT, et al: Immunostaining of neurofibrillary tangles in Alzheimer's senile dementia with a neurofilament antiserum. J Neurosci 2(1):113-119, 1982

18. Anderton BH, Breinburg D, Downes MJ, et al: Monoclonal antibodies show that neurofibrillary tangles and neurofilaments share antigenic determinants. Nature 298:84-86, 1982
19. Gambetti P, Shecket G, Ghetti B, et al: Neurofibrillary changes in human brain: An immunocytochemical study with a neurofilament antiserum. J Neuropathol Exp Neurol 42(1):69-79, 1983
20. Rasool CG, Abraham C, Anderton BH, et al: Alzheimer's disease: Immunoreactivity of neurofibrillary tangles with anti-neurofilament and anti-paired helical filament antibodies. Brain Res 310:249-260, 1984
21. Perry G, Rizzuto N, Autilio-Gambetti L, et al: Paired helical filaments from Alzheimer disease patients contain cytoskeletal components. Proc Nat'l Acad Sci USA 82:3916-3920, 1985
22. Elovaara I, Paetau A, Lehto V-P, et al: Immunocytochemical studies of Alzheimer neuronal perikarya with intermediate filament antisera. J Neurol Sci 62:315-326, 1983
23. Yen S-H, Gaskin F, Fu SM: Neurofibrillary tangles in senile dementia of the Alzheimer type share an antigenic determinant with intermediate filaments of the vimentin class. Am J Pathol 113(3):373-381, 1983
24. Kosik KS, Duffy LK, Dowling MM, et al: Microtubule-associated protein 2: Monoclonal antibodies demonstrate the selective incorporation of certain epitopes into Alzheimer neurofibrillary tangles. Proc Nat'l Acad Sci USA 81:7941-7945, 1984
25. Yen S-H, Dickson DW, Crowe A, et al: Alzheimer neurofibrillary tangles contain unique epitopes and epitopes in common with the heat-stable microtubule associated proteins tau and Map 2. Amer J Pathol 126(1):81-91, 1987
26. Ihara Y, Abraham C, Selkoe DJ: Antibodies to paired helical filaments in Alzheimer's disease do not recognize normal brain proteins. Nature 304:727-730, 1983
27. Wood JG, Mirra SS, Pollock NJ, et al: Neurofibrillary tangles of Alzheimer disease share antigenic determinants with the axonal microtubule-associated protein tau. Proc Nat'l Acad Sci USA 83:4040-4043, 1986
28. Kosik KS, Joachim CL, Selkoe DJ: Microtubule-associated protein tau is a major antigenic component of paired helical filaments in Alzheimer disease. Proc Nat'l Acad Sci USA 83: 4044-4048, 1986
29. Grundke-Iqbal I, Iqbal K, Tung Y-C, et al: Abnormal phosphorylation of the microtubule-associated protein tau in Alzheimer cytoskeletal pathology. Proc Nat'l Acad Sci USA 83:4913-4917, 1986
30. Binder LI, Frankfurter A, Rebhun LI: The distribution of tau in the mammalian central nervous system. J Cell Biol 101:1371-1378, 1985
31. Pollock NJ, Binder LI, Mirra SS, et al: Alzheimer's disease: A model for study of cytoskeletal segregation as a determinant of neuronal form and function, in Lasek RJ (ed): Intrinsic Determinants of Neuronal Form, New York, Alan R Liss Inc (In press)
32. Munzo-Garcia D, Ludwin SK: Classic and generalized variants of Pick's disease: A clinicopathological, ultrastructural, and immunocytochemical comparative study. Ann Neurol 16(4):467-480, 1984
33. Tellez-Nagel I, Wisniewski HM: Ultrastructure of neurofibrillary tangles in Steele-Richardson-Olszewski syndrome. Arch Neurol 29:324-327, 1973
34. Tomonaga M: Ultrastructure of neurofibrillary tangles in progressive supranuclear palsy. Acta Neuropath (Berl) 37:177-181, 1977
35. Dickson DW, Kress Y, Crowe A, et al: Monoclonal antibodies to Alzheimer neurofibrillary tangles 2. Demonstration of a common antigenic determinant between ANT and neurofibrillary degeneration in progressive supranuclear palsy. Am J Pathol 120(2):292-303, 1985

36. Probst A, Anderton BH, Ulrich J, et al: Pick's disease: An immunocytochemical study of neuronal changes. Acta Neuropathol (Berl) 60:175-182, 1983
37. Rasool CG, Selkoe DJ: Sharing of specific antigens by degenerating neurons in Pick's disease and Alzheimer's disease. N Engl J Med 312(11):700-705, 1985
38. Goldman JE, Yen S-H, Chio F-C, et al: Lewy bodies of Parkinson's disease contain neurofilament antigens. Sci 221:1082-1084, 1983
39. Pappolla, MA: Lewy bodies of Parkinson's disease. Arch Pathol Lab Med 110:1160-1163, 1986
40. Ditter SM, Mirra SS: Neuropathological and clinical features of Parkinson's disease in Alzheimer's disease patients. Neurol (In press)
41. Hirano A: Pathology of amyotrophic lateral sclerosis. NINDB Monograph No. 2 in Gajdusek DC, Gibbs CJ Jr, Alpers M (eds): Slow, Latent and Temperate Virus Infections. US Department of Health, Education and Welfare, National Institute of Health, National Institute of Neurological Diseases and Blindness, Public Health Service publication, Washington DC, 1965 p.23
42. Goldman JE: The association of actin with Hirano bodies. J Neuropathol Exp Neurol 42(2):146-152, 1983
43. Galloway PG, Perry G, Gambetti P: Hirano body filaments contain actin and actin associated proteins. J Neuropathol Exp Neurol 46(2):185-199, 1987
44. Ogata J, Budzilovich GN, Cravioto H: A study of rod-like structures (Hirano bodies) in 240 normal and pathological brains. Acta Neuropathol (Berl) 21:61-67, 1972
45. Gibson PH, Tomlinson BE: Numbers of Hirano bodies in the hippocampus of normal and demented people with Alzheimer's disease. J Neurol Sci 33:199-206, 1977
46. Rewcastle NB, Ball MJ: Electron microscopic structure of the "inclusion bodies" in Pick's disease. Neurol 18:1205-1213, 1968
47. Schochet SS Jr, Lampert PW, Lindenberg R: Fine structure of the Pick and Hirano bodies in a case of Pick's disease. Acta Neuropathol 11:330-337, 1968
48. Hirano A, Malamud N, Elizan TS, et al: Amyotrophic lateral sclerosis and parkinsonism-dementia complex on Guam: Further pathological studies. Arch Neurol 15:35-51, 1966
49. Pollock NJ, Mirra SS, Binder LI, et al: Filamentous aggregates in Pick's disease, progressive supranuclear palsy, and Alzheimer's disease share antigenic determinants with microtubule-associated protein, tau. Lancet November 22:1211, 1986
50. Weingarten MD, Lockwood AH, Hwo SY, et.al: A protein factor essential for microtubule assembly. Proc Nat'l Acad Sci USA 72:1858-1862, 1975
51. Galloway PG, Perry G, Kosik KS, et al: Hirano bodies contain tau protein. Brain Res 403:337-340, 1987

CYTOSKELETAL ABNORMALITIES IN PARKINSON'S DISEASE

James E. Goldman

Department of Pathology
Albert Einstein College of Medicine
Bronx, NY

Alterations in the neuronal cytoskeleton are major features of many neurodegenerative diseases. In Parkinson's Disease, a characteristic pathological change is the Lewy body. Lewy bodies are intraneuronal, cytoplasmic inclusions. They were first described in large neurons of the substantia innominata and dorsal motor nucleus of the vagus (1). We now consider the catecholaminergic nuclei, the substantia nigra and locus ceruleus, to be the typical sites of Lewy body formation in Parkinson's disease. Nevertheless, careful neuropathological studies have demonstrated Lewy bodies in a large number of CNS areas, including many brain stem nuclei, hypothalamus, neocortex, and peripheral ganglia (2-5). Lewy bodies in these different areas show many similarities, but some morphological differences (see below).

The first studies to link Lewy bodies to cytoskeletal elements were electron microscopic observations (4,6-8). The inclusions contain large numbers of filaments, 6-8nm in diameter. In brain stem neurons, the filaments tend to be oriented radially around a core composed of vesicular and granular material. In cortical neurons, a radial arrangement is not observed, and filaments appear more randomly organized into a round mass, admixed with granular and membranous material (8,9). Lewy bodies are not membrane bound, nor do they seem to be associated with specific intracytoplasmic membrane systems. The neuromelanin granules of pigmented brain stem neurons are not part of the inclusion, but rather appear pushed to the side by the filamentous masses. In most locations, Lewy bodies are to be confined to the neuronal perikaryon. In peripheral ganglion cells, however, inclusions are found in proximal processes.

At the time of the first ultrastructural observations, far less was known about the specific molecular constituents of the neuronal cytoskeleton than is known today. The present knowledge of the major neuronal filamentous proteins, and the generation of antibodies to them, has made possible some insights into the nature of Lewy bodies. Our current understanding of the composition of Lewy bodies has been derived entirely from immunocytochemical studies, because it has not yet been possible to isolate the inclusions for direct biochemical analysis. Antibody reactions can be summarized as follows.

Several studies have demonstrated binding of anti-neurofilament antibodies to Lewy bodies, indicating the presence of neurofilament antigens (10-14). The size of the filaments places them in the range of intermediate filaments, of which neurofilaments are the neuronal-specific class. Antibodies include polyclonal rabbit antisera, which react with all three neurofilament proteins or are monospecific, and mouse monoclonal antibodies, which react mainly with the 200kd and 150kd peptides Thus, there is evidence that all three neurofilament components can be found in Lewy bodies. Whether they are organized in a normal fashion has not yet been determined. Some antibodies give immunocytochemical reactions over the entire inclusion, while others react more with the periphery and less with the core. While this may be an artifact, related to how the inclusion is sectioned, it also may reflect how different neurofilament antigens are distributed within the inclusion. Monoclonal antibodies tend to stain the periphery, while we have seen both peripheral and overall staining with rabbit antisera. Some of the neurofilament epitopes within Lewy bodies are phosphorylated. This has been inferred from positive reactions with monoclonal antibodies, the binding of which to neurofilaments can be prevented by treatment of the proteins or the tissues with phosphatase (13). Whether phosphorylation of neurofilaments is of primary importance in the formation of Lewy bodies or is a secondary consequence of other changes is not yet clear (see below).

In collaboration with Dr. Richard Vallee, we have found immunocytochemical reactions of Lewy bodies with antibodies to high molecular weight microtubule-associated proteins (MAPs) (Figure 1). Antibodies to MAP1 and to MAP2 bind to the inclusions, and the reaction is observed in both brain stem and cortical Lewy bodies. A positive reaction with anti-tubulin antibodies has been reported (14), but our experience has been that anti-tubulin antibodies bind only weakly to Lewy bodies. Normal appearing microtubules are not part of Lewy body structure. These observations raise the possibility that high molecular weight MAPs become associated with Lewy bodies because they bind to other constituents, possibly neurofilaments or perhaps tubulin (see below).

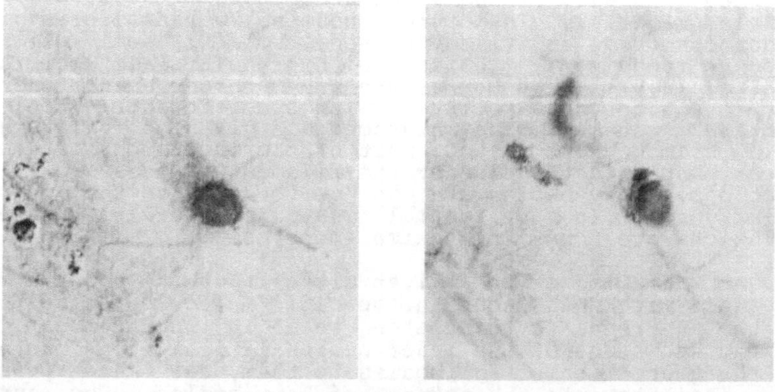

Figure 1: Reactions of anti-MAP1a antibody with Lewy bodies in the substantia nigra. Antibody binding was visualized using the peroxidase-anti-peroxidase technique. Round, immunoreactive profiles are present in cell perikarya. Perikaryal and dendritic staining is also evident. X310.

The filaments of Lewy bodies are morphologically different from the paired helical filaments of neurofibrillary tangles and the 15-20nm straight filaments seen in tangles and in Pick bodies. Some antibodies that bind to tangles also react with Lewy bodies, while others do not (10-12, 14). The specific epitopes that are recognized by these various antibodies in Lewy bodies and tangles have not yet been identified. It is likely that mechanisms of formation of Lewy bodies are different from that generating tangles. If both types of inclusions contain the same kinds of cytoskeletal proteins (neurofilaments, for example), albeit altered in different ways, then one might expect some cross-reactive antibodies as well as non-cross-reactive ones. The dissimilarities between Lewy bodies and paired helical filaments does not reflect intrinsic differences in the ability of various neurons to form cytoskeletal abnormalities, since nigral neurons are well known to form tangles in several disorders.

Contractile proteins, which are associated with Hirano bodies, do not form part of the Lewy body, at least judging from immunocytochemical studies (14 and our unpublished observations.

The following table summarizes immunocytochemical studies with antibodies to cytoskeletal proteins

Table 1

Antigen	Binding to LBs
Neurofilaments	+
MAP1a	+
MAP2	+
Tubulin	±
PHF	±
Actin	−
Myosin	−
Tropomyosin	−

It is important to note that Lewy bodies include components other than cytoskeletal elements. That has been clear from ultrastructural observations. Antibodies to tyrosine hydroxylase bind to the inclusions (15). Monoclonal antibodies, generated to Parkinson's disease nigra, bind to Lewy bodies (16). The antigens appear to be proteins, but have not been fully characterized. Whether non-cytoskeletal elements play a critical role in the genesis and organization of Lewy bodies, or whether cytoskeletal disorganization is primary, with other elements following the filamentous abnormality is a critical question which has not been answered. Typical Lewy bodies have not been generated experimentally. Furthermore, early stages of Lewy body formation have not been reported, although Forno has suggested that the cytoplasmic clearing seen in nigral and locus ceruleus neurons in Parkinson's may represent an early cytoplasmic event.

A number of possible mechanisms could account for Lewy body formation. These will be considered separately, but are not mutually exclusive.

The accumulation of filaments within cell perikarya suggests alterations in transport or in the export of filaments from the cell body, or alterations in turnover. Decreased transport of neurofilaments is consistent with finding Lewy bodies in proximal processes in peripheral ganglia. It is common to see many axonal spheroids in the nigra in Parkinson's disease. It has not been possible, however, to discern the neurons of origin. Alterations in turnover would add to accumulations. Intermediate filaments are substrates of calcium-activated proteases in vitro (17). Such proteases may be active at synaptic terminals, an idea consistent with the fact that neurofilaments are transported along axons, but are not prominent organelles in synapses. Neurofilaments can be altered by proteolysis during transport in some axonal systems (18). Mechanisms and degree of turnover of filaments in neuronal cell bodies is not known.

Post-translational modifications of filaments could account for alterations in export and turnover. The most popular idea currently has to do with phosphorylation of filaments. Antibodies that bind to phosphatase-sensitive epitopes react with Lewy bodies. These antibodies do not normally react well with neuronal perikarya, binding largely to axons. These observations have raised the idea that phosphorylation of neurofilaments is important in generating the abnormality. Binding of these antibodies to neuronal perikarya is an interesting phenomenon, but it is not specific for Parkinson's disease. It can be seen in a variety of conditions, including motor neuron disease, aluminum neurotoxicity, and experimental chromatolysis (19-21). Neurofilaments do accumulate abnormally in perikarya and proximal axons in these disorders, but the typical Lewy body structures are not formed. Other forms of amino acid derivitization could have a number of effects, such as slowing or preventing turnover by changing amino-termini (22), or preventing the normal interactions between cytoskeletal systems (see below). The idea that post-translational modifications of cytoskeletal elements may alter the ability of intracellular proteolytic mechanisms to degrade the proteins is important to consider. The apparent stability of cytoskeletal abnormalities in degenerative diseases may well be directly related to the neuron's inability to recognize and degrade altered cytoskeletal substrates.

Changes in the organization of microtubule systems have consequences for intermediate filament organization. Thus, depolymerizing microtubules in cultured cells leads to a coiling of intermediate filaments (23). The disease giant axonal neuropathy may indeed be a disorder of connections between microtubular and intermediate filament systems, since fibroblasts from patients with this disease show the characteristic filament coils (24). Accumulations of intermediate filaments are evident in many cell types, including neurons. Lewy bodies have not been described. Intoxications with drugs such as b,b'-iminodiproprionitrile also cause a dissociation of microtubules from neurofilaments, and result in the loss of axonal transport of neurofilaments (25-27). Thus, abnormalities in one cytoskeletal system can have profound consequences for the organization of other filamentous proteins. Finding high molecular weight MAPs associated with the intermediate filament accumulations in GAN and in toxic states demonstrates that MAPs can interact with intermediate filaments (23,25). Furthermore, in vitro interactions between MAPs and neurofilaments has been demonstrated (28). If MAPs are included in Lewy bodies by virtue of their association with neurofilaments, it might be useful to consider the possibility that microtubule systems are disrupted in Parkinson's disease.

Parkinson's disease induced by MPTP involves the death of neurons in the substantia nigra (29-31). A recent study reported the formation of acidophilic inclusions in nigral neurons in monkeys given MPTP (31). Whether these inclusions bear similarities to Lewy bodies remains to be seen, but their formation is worth study. For example, the active compound in MPTP intoxication appears to be the oxidized form, 1-methyl-4-phenylpyridinium ion. One could postulate that a direct binding of such a cationic toxin to neurofilaments might alter the interactions of filaments with other cytoskeletal systems, as has been proposed for other toxins (32), thus segregating neurofilaments and halting their export from the cell body and resulting in a disorganized accumulation. Chemical modification of filaments may have the added effect of altering turnover rates of the proteins. Examination of cytoskeletal abnormalities in such a model may lead to important insights into the genesis of neurofilamentous pathology.

REFERENCES

1. Lewy FH: Paralysis agitans. I. Pathologische Anatomie, in Lewandowsky M (ed): Handbuch der Neurologie. Springer, Berlin, 1912, pp920-933.

2. Jager WA Den Hartog and Bethlem J: The distribution of Lewy bodies in the central and autonomic nervous system in idiopathic paralysis agitans. J Neurol Neurosurg Psychiatry 23:283-290, 1960.

3. Ohama E and Ikuta F: Parkinson's disease: distribution of Lewy bodies and monoamine neuron systems. Acta Neuropathol (Berl) 34:311-319, 1976.

4. Forno LS and Norville RL: Ultrastructure of Lewy bodies in the stellate ganglion. Acta Neuropathol (Berl) 34:183-197, 1976.

5. Kosaka K, Yoshimura M, Iheda K, and Budka H: Diffuse type of Lewy body disease: progressive dementia with abundant cortical Lewy bodies and senile changes of varying degree-a new disease? Clin Neuropath 3:185-192, 1984.

6. Duffy PE and Tennyson VM: Phase and electron microscopic observations of Lewy bodies and melanin granules in the substantia nigra and locus ceruleus in Parkinson's disease. J Neuropathol Exp Neurol 24:398-414, 1965.

7. Roy S and Wolman L: Ultrastructural observations in parkinsonism. J Pathol 99:39-44, 1969.

8. Kosaka K: Lewy bodies in the cerebral cortex. Report of three cases. Acta Neuropathol (Berl) 42:127-134, 1978.

9. Yen S-H, Dickson D, Peterson C, and Goldman JE: Cytoskeletal abnormalities in neurodegenerative diseases. Prog Neuropathol 6:63-90, 1986.

10. Goldman JE, Yen S-H, Chiu FC and Peress N: Lewy bodies of Parkinson's disease contain neurofilament antigens. Science 221:1082-1084, 1983.

11. Nakazato Y, Sasaki A, Hirato J, and Ishida Y: Immunohistochemical localization of neurofilament protein in neuronal degenerations. Acta Neuropathol (Berl) 64:30-36, 1984.

12. Kahn J, Anderton BH, Gibb WRG, Lees AJ, Wells FR, and Marsden CD: Neuronal filaments in Alzheimer's, Pick's and Parkinson's diseases. New Engl J Med 313:520-521, 1985.

13. Forno LS, Sternberger LA, Sternberger NH, Strefling AM, Swanson K and Eng LF: Reaction of Lewy bodies with antibodies to phosphorylated and non-phosphorylated neurofilaments. Neurosci Lett 64:253-258, 1986.

14. Galloway P, Perry G, Grundke-Iqbal I, Autilio-Gambetti L, and Gambetti P: Lewy bodies share epitopes with neurofilaments, tubulin, and neurofibrillary tangles. Xth Internat Congress Neuropathol Abstracts p298, 1986.

15. Nakashima S and Ikuta F: Tyrosine hydroxylase protein in Lewy bodies of parkinsonian and senile brains. J Neurol Sci 66:91-96, 1984.

16. Hirsch E, Ruberg, Dardenne M, Partier M-M, Javoy-Agid R, Bach J-F, and Agid Y: Monoclonal antibodies raised against Lewy bodies in brains from subjects with Parkinson's disease. Brain Res 345:374-378, 1985.

17. Schlaepfer WW: Nature of mammalian neurofilaments and their breakdown by calcium. Prog Neuropathol 3:101-123, 1979.

18. Nixon RA: Proteolysis of neurofilaments, in Marotta CA (ed): Neurofilaments. Minneapolis, University of Minneapolis Press, 1983, pp117-154.

19. Troncoso J, Sternberger NH, Sternberger LA, Hoffman PN and Price DL: Immunocytochemical studies of neurofilament antigens in the neurofibrillary pathology induced by aluminum. Brain Res 364:295-300, 1986.

20. Wiley CA, Love S, Skoglund RR, and Lampert PW: Infantile neurodegenerative disease with neuronal accumulation of phosphorylated neurofilaments. Acta Neuropathol (Berl) 72:369-376, 1987.

21. Greene C, Munoz-Garcia D, Perl D and Pendlebury W: Accumulation of phosphorylated neurofilaments in the anterior horn motor neurons of ALS patients. J Neuropathol Exp Neurol 45:332, 1986 (abstract).

22. Bachmair A, Finley D and Varshavsky: In vivo half-life of a protein is a function of its amino-terminal residue. Science 234:179-186, 1986.

23. Bloom GS and Vallee RB: Association of microtubule-associated protein 2 (MAP2) with microtubules and intermediate filaments in cultured brain cells. J Cell Biol 96:1523-1531, 1983.

24. Pena SDJ: Giant axonal neuropathy: intermediate filament aggregates in cultured skin fibroblasts. Neurology 31:1470-1473, 1981.

25. Papasozomenos SCH, Binder LI, Bender PK, Payne MR: Microtubule-associated protein 2 within axons of spinal motor neurons: associations with microtubules and neurofilaments in normal and B-B'-iminodipropionitrile-treated axons. J Cell Biol 100:74-85, 1985.

26. Griffin JW, Fahnestock KE, Price DL and Hoffman PN: Microtubule-neurofilament segregation produced by beta-beta'-iminodipropionitrile: evidence for the assocation of fast axonal transport with microtubules. J Neurosci 3:557-566, 1983.

27. Griffin JW, Hoffman PN, Clark AW, Carrol PT and Price DL: Slow axonal transport of neurofilament proteins: impairment by B-B'-iminodipropionitrile administration. Science 202:633-638, 1978.

28. Leterrier JF, Wong J, Liem RHK and Shelanski ML: Promotion of microtubules assembly by neurofilament-associated microtubule-associated proteins. J Neurochem 43:1385-1391, 1984.

29. Ballard PA, Tetrud JW and Langston JW: Permanent human parkinsonism due to 1-methyl-4-phenyl-1,2,3,6-tetrahydropyridine (MPTP): Seven cases. Neurology 35:949-956, 1985.

30. Burns RS, Chiueh CC, Markey SP et al.: A primate model of parkinsonism: selective destruction of dopaminergic neurons in the pars compacta of the substantia nigra by n-methyl-4-phenyl-1,2,3,6 tetrahydropyridine. Proc Natl Acad Sci USA 80:4546-4550, 1983.

31. Forno LS, Langston JW, DeLanney LE, Irwin I, and Ricaurte GA: Locus ceruleus lesions and eosinophilic inclusions in MPTP-treated monkeys. Ann Neurol 20:449-455, 1986.

32. Sayre LM, Autilio-Gambetti L and Gambetti P: Pathogenesis of experimental giant neurofilamentous axonopathies: a unified hypothesis based on chemical modification of neurofilaments. Brain Res Reviews 10:69-83, 1985.

MICROFILAMENT INVOLVEMENT IN HIRANO BODY FORMATION

Pamela G. Galloway and George Perry

Case Western Reserve University
Cleveland, Ohio

The neuron, like many other types of cells, contains three major morphologically distinct filamentous systems: microfilaments, intermediate filaments and microtubules.[1] The role of these cytoskeletal components in the neuron has been unclear until recently. These discoveries have implicated the cytoskeleton in many cell functions, including internal organization.[2-5] Microfilaments in neurons have mainly been studied in relation to axonal transport.[6-9] By light microscopy, actin filaments have been demonstrated in neurites,[10-13] growth cones of avian dorsal root ganglia cells[14-18] and in microspikes of murine neuroblastoma cells.[10,11] In situ organization of actin filaments in dendritic spines, dendrites and axon terminals has been demonstrated[19] using heavy meromyosin decoration.[20]

Microfilaments, and other components of the neuronal cytoskeleton, have been implicated in degenerative diseases of the central nervous system. These diseases are characterized morphologically by loss of neurons from certain areas of the brain, and by inclusion bodies within neurons. Alzheimer's and Pick's diseases and progressive supranuclear palsy are characterized by intracytoplasmic inclusions which contain components of the neuronal cytoskeleton.[21-25]

The Hirano body is another such inclusion. It is an eosinophilic, usually rod-shaped structure originally described in a patient with Parkinsonism dementia complex of Guam.[26] It is commonly present in neurons of Sommer's sector of the hippocampus[27] and in the underlying stratum lacunosum.[28] Less commonly it is located in spinal cord[29], cerebellum,[30] glial cells,[31] cerebral cortex,[32] Schwann cells,[33] Purkinje cells,[34,35] inflammatory cells,[36] oligodendroglia,[37] and extraocular muscles.[38] The Hirano body can be seen in hippocampal neurons of non-demented old individuals[39-41] and in increased numbers in patients with Alzheimer's disease.[42] It is also seen in Pick's disease,[43] sporadic motor neuron disease,[27] hepatic encephalopathy,[36,44] the ataxic form of Jakob-Creutzfeld disease,[45] in axons of peripheral nerves in patients with progressive external ophthalmoplegia,[46] and in Seitelberger's connatal form of Pelizaeus-Merzbacher disease.[47] Hirano bodies are also seen in experimental animals, including squirrel monkey dorsal horn,[48] in Schwann cells of rat

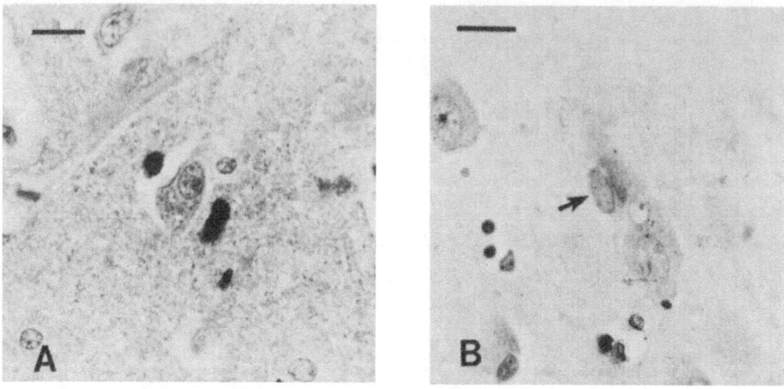

Fig. 1. A(left) Two Hirano bodies adjacent to a hippocampal neuron immunostained with a smooth muscle tropomyosin antibody using the peroxidase antiperoxidase (PAP) technique. B(right). Immunostaining of the Hirano body (arrow) is removed after adsorption with tropomyosin. (Scale bar= 58 microns).

peripheral nerve,[49-52] spinal roots of a hamster mutant with hind leg paralysis,[33] in mice with experimental scrapie,[53] chimpanzees with experimental kuru,[30] peripheral nerves of the spontaneously diabetic B-B Wistar rat,[54] in brains of ferrets inoculated with subacute sclerosing panencephalitis,[55] and in the brindled mouse.[56-59]

Immunocytochemical analyses have implicated microfilaments in Hirano body formation. The major component of microfilaments is actin, a 43,000 kilodalton (kD) protein.[60] In cells actin can either be present as monomers, i.e., in the globular or G state, or as filaments, i.e. in the F state.[61] Actin molecules assemble and disassemble to form three dimensional structures to meet the functional requirements of the cell. This assembly and dissembly of actin is controlled by a group of proteins called actin associated proteins.[61-63] These proteins are classified into groups, including proteins that predominantly sequester actin monomers, proteins which block the ends of actin filaments, actin filament cross-linking proteins, and proteins that bind to the sides of actin filaments.[62]

Immunocytochemical studies have shown that Hirano bodies share epitopes with actin[28,64] and actin associated proteins including vinculin, alpha-actinin and tropomyosin (Fig. 1).[64] The epitopes are distributed evenly throughout the body, as was demonstrated by immunoelectron microscopy (Fig. 2).[64]

At least nine different isoforms of tropomyosin have been described.[65] The molecular weights of the different forms vary with cell type, and even within the same cell, in the range of 29-39 kD.[62] Some tropomyosin isoforms have been isolated from brain.[66,67] Antibodies raised to certain isoforms cross react with other isoforms.[68] Tropomyosin epitopes in Hirano bodies immunoreacted with antibodies to both platelet and smooth muscle[64] tropomyosin; but not with antibodies to skeletal muscle[69] or to brain tropomyosin.[70]

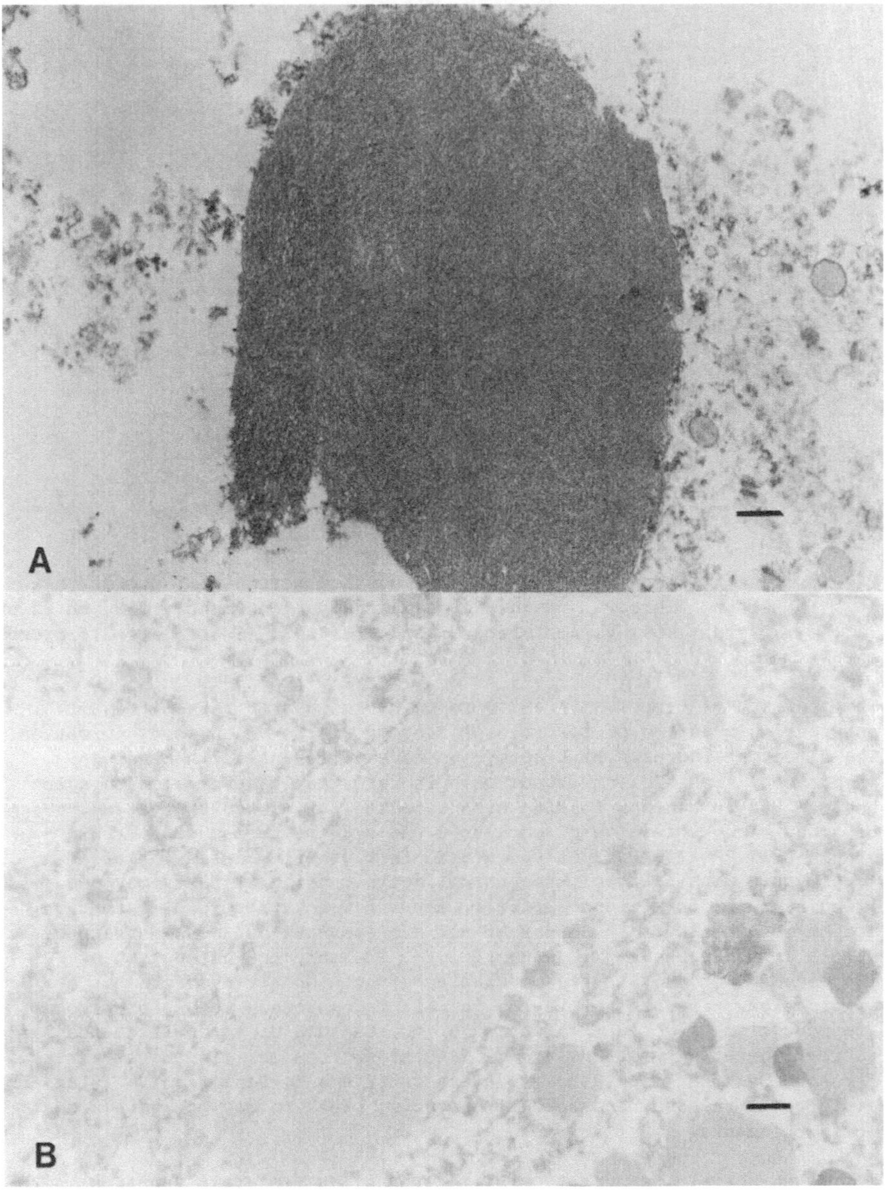

Fig. 2. A(upper) Hirano body viewed at the electron microscopic level immunostained with a platelet tropomyosin antibody. B(lower) The Hirano body immunostained with rabbit pre-immune serum as primary antibody is unstained. Sections were not contrasted with uranyl acetate and lead citrate. (Scale bar= 0.5 micron.)

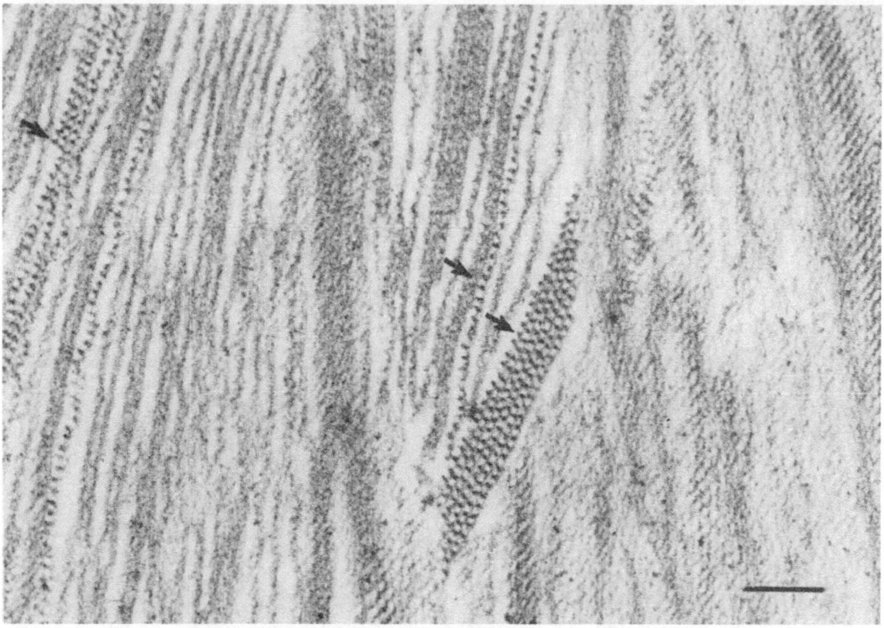

Fig. 3. The filaments are in a Hirano body are arranged as "beads on a string" (arrows) or in a crosshatched arrangement (right third of field). Uranyl acetate and lead citrate. (Scale bar= 0.5 micron.)

The presence of tropomyosin epitopes in Hirano bodies raises the interesting question of the role of tropomyosin in Hirano body formation. The different isoforms of tropomyosin have different actin binding properties.[71] Different tropomyosin isoforms may be expressed in aging neurons and in neurons forming Hirano bodies. Evidence indicates there may be genetic co-expression of more than one type of tropomyosin within a single cell.[65] It might be speculated that genetic expression of a tropomyosin with a higher than normal actin binding affinity may occur in aging neurons; this higher affinity may initiate Hirano body formation. Tropomyosin binds to the sides of actin filaments, but does not appear to have a direct cross-linking function.[62] The only known function of tropomyosin, other than its modulating effects on actin-myosin interaction,[72] is its ability to inhibit incompletely the spontaneous fragmentation of actin filaments[72] or the rupture of actin filaments by actin-fragmenting proteins.[74,75] These functions may play a role in Hirano body formation, since spontaneous fragmentation of actin filaments would be diminished, and actin fragmenting proteins may not gain access to actin filaments.

Alpha-actinin is an actin filament bundling protein. Proteins in this group stablize the alignment of actin filaments into side-by-side aggregates. Alpha-actinin is an asymmetrical homodimer with a molecular weight of 105 kD, which has been isolated from brain cells.[76] Non-muscle alpha-actinins are calcium binding proteins, a feature which distinguishes them from muscle alpha-actinin. Calcium binding inhibits the ability of alpha-actinin to interact with actin.[61] As non-muscle alpha-actinin binds calcium, its binding affinity for actin diminishes.[76,77] The ability of alpha-actinin to bind and cross-link actin filaments into a side-by-side configuration occurs maximally at low temperatures in vitro, at least in the case of some cytoplasmic alpha-actinins.[61] Whether the same property is also true of neuronal alpha-actinin is unknown.

The presence of vinculin epitopes is also intriguing. Vinculin was formerly classified as an actin bundling protein, based on its location in areas where actin filaments insert into the plasma membrane.[78] This concept was supported by subsequent studies showing that vinculin preparations obtained according to published procedures[79] affect assembly of actin filaments and lower the viscosity of F-actin measured by low shear viscometric techniques.[80-84] However, it is known that contaminants in the vinculin preparations are responsible for practically all of the filament bundling activities previously attributed to vinculin.[85] Studies have shown that vinculin of higher purity than those used in previous studies has little or no effect on actin assembly and on the low shear viscosity of F-actin solutions.[85-89] Given the apparent lack of interaction between vinculin and actin, it is intriguing that vinculin epitopes are present in Hirano bodies. It can be speculated that the impurity may also be incorporated into the Hirano body.

With electron microscopic examination, Hirano bodies show a paracrystalline structure made of sheets of 6-10 nm parallel filaments. The filaments of adjacent sheets are arranged at oblique angles.[27,29] The filaments in Hirano bodies can either appear as a cross-hatched arrangement or as "beads on a string", (Fig 3) depending on the angle of view. Immunostaining at the electron microscopic level has shown that filaments in both arrangements are immunostained (Fig. 4) when antibodies which immunostain Hirano bodies at the light microscopic level are used.[64] The crossover periodicity of the filaments is regular and ranges from 20 to 25 nm.[58] Various interpretations of this arrangement have been suggested. One interpretation[90] is that the "beads" are granular ribosomal material arranged on the "strings". Another interpretation[91] using digital processing on electron micrographs is that they are composed of helical strands. However, other evidence[64] confirms the purely filamentous nature of Hirano bodies. High voltage electron microscopy performed on 1.0 micron and 0.5 micron sections showed that the "beads" on the "bead on a string" arrangement were transformed into filaments when viewed at the proper orientation (Fig 5). The section thickness afforded a greater representation of the change from beads to filaments for a given angle of rotation. Because the "beads" in the "bead on a string" arrangement were visualized as filaments after tilting 1.0 and 0.5 micron sections, this finding indicates that Hirano bodies are made entirely of superimposed filaments, and do not contain granular structures. In this figure, it can be seen that the beads are in fact cross sections of filaments rather than granules.

When frozen sections of hippocampus from Alzheimer's disease patients were treated with rhodamine and fluorescein labeled phalloidin, probes with unique affinity for F-actin,[92] Hirano bodies were stained.[64] This indicates that actin in Hirano bodies is in the F state.

Although Hirano bodies appear to be derived primarily from neuronal microfilaments, the neurofibrillary tangle, and its main constituent, paired helical filaments[93], do not show immunologic cross-reactivity with components of microfilaments.[64] Antibodies raised against sodium dodecyl sulfate extracted paired helical filaments show reactivity with the microtubule associated protein, tau.[94-96] Several antibodies against the 200 kilodalton (kD) and 145 kD neurofilament subunits[21,22,97] and microtubule associated protein 2 (MAP2),[98] and the microtubule associated protein tau are also immunoreactive with neurofibrillary tangles.[94-96,99,100] To determine a possible relationship between

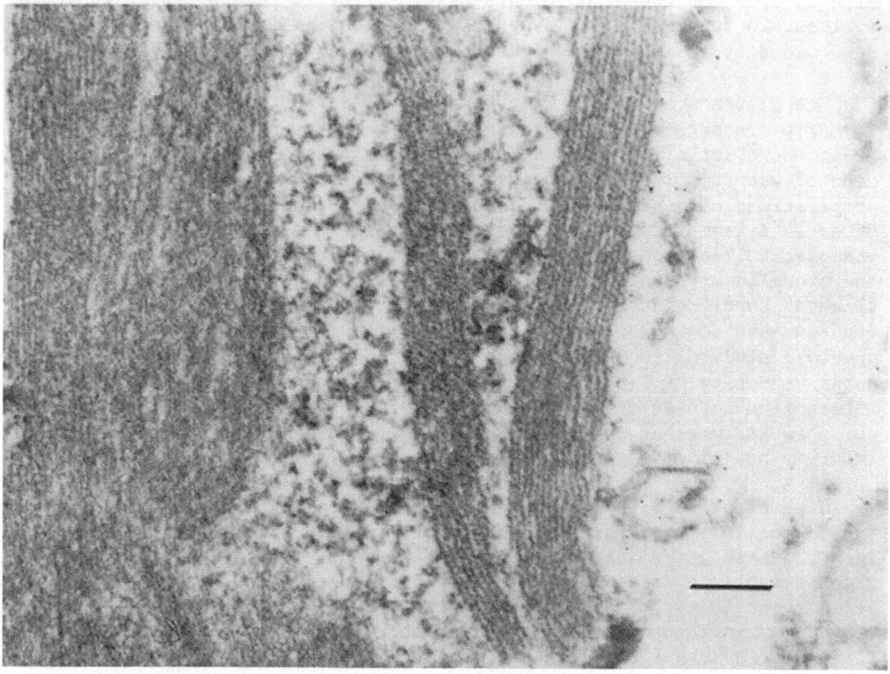

Fig. 4. The platelet tropomyosin antibody reacts intensely in areas both with the "bead on a string" arrangement of filaments (right half of Hirano body) and the crosshatched arrangement (left half). The material between the large groups of Hirano body filaments is osmiophilic material of undetermined composition, which is seen in some Hirano bodies. (Scale bar= .31 micron.)

neurofibrillary tangles and Hirano bodies, Hirano bodies were immunostained with a monoclonal (5E2) antibody[101] and a polyclonal antibody[102] to tau protein. Hirano bodies were immunostained by the polyclonal antibody and by 5E2. In contrast to the results obtained using antibodies to actin, tropomyosin, vinculin and alpha-actinin, only about 20% of Hirano bodies in a section were immunostained. Tau is a microtubule associated protein[103] which can also serve as an actin cross-linking protein.[60-63] Fractions enriched in both MAP2 and tau have been shown to gel actin at relatively high concentrations.[61] Actin filaments incubated with high concentrations of tau form bundles visible in the electron microscope.[104] Using antibodies to MAP1a and 2,[105] the presence of these epitopes has been shown in Hirano bodies.[106] Since MAPs localize to microtubules, and not to cortical networks or bundles of actin in cells,[61] the functional significance of actin cross-linking by MAPs is not clear. It is possible that they become complexed under pathological conditions.[106]

As yet, it is not clear what role MAPs play in Hirano body and neurofibrillary tangle formation. Although it is possible that Hirano bodies are mainly derived from microfilaments and neurofibrillary tangles derive largely from neurofilaments and microtubules, it is possible that the underlying mechanisms of their formation may be related.

This work was supported by NIH Grant AG00795.

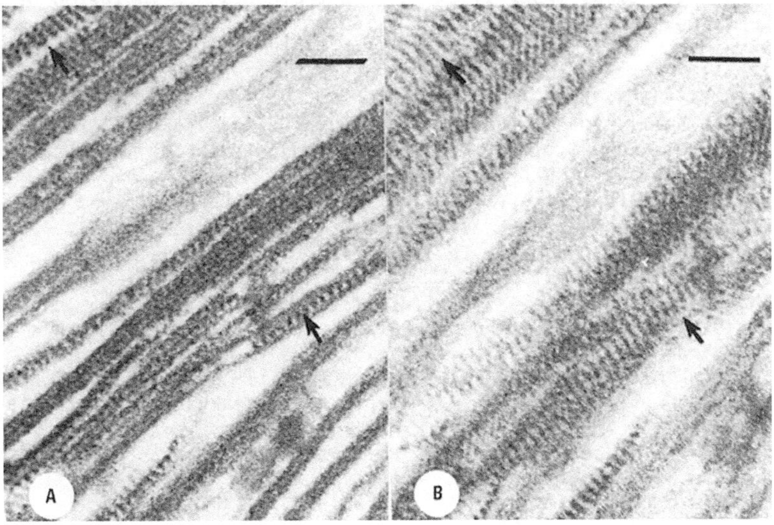

Fig. 5. Electron micrograph of a 0.5 micron plastic embedded section examined at one million electron volts. The specimen on the right is positioned at -10° and on the left at +10°. The "beads" (arrows) change to filaments (arrows). (Scale bar= .24 micron.)

REFERENCES

1. Fifkova E: Actin in the nervous system. Brain Res Rev, 9:187-215, 1985.
2. Ellisman MH, Porter KR: Microtrabecular structure of the axoplasmic matrix: visualization of cross-linking structures and their distribution. J Cell Biol, 87:464-479, 1980.
3. Ellisman MH, Porter KR: Introduction to the cytoskeleton, in: "Neurofilaments", C.A. Marotta, ed. University of Minnesota Press, Minneapolis, 1983.
4. Schliwa M, Pryzwansky KB, Van Blerkom J: Implications of cytoskeletal interactions for cellular architecture and behavior. Phil Trans Roy Soc B, 299:199-205, 1982.
5. Schliwa M, Van Blerkom J, Pryzwansky KB: Structural organization of the cytoplasm. Cold Spring Harbor Symp Quant Biol, 46:51-67, 1982.
6. Black M, Lasek R: Axonal transport of actin: slow component b is the principal source of actin for the axon. Brain Res, 171:401-413, 1979.
7. Black M, Lasek R: Slow components of axonal transport: two cytoskeletal networks. J Cell Biol, 86:616-623, 1980.
8. Grafstein B, Forman DS: Intracellular transport in neurons. Ann Rev Physiol, 60:1167-1283, 1980.
9. Schwartz JH: Axonal transport: components, mechanisms, and specificity. Ann Rev Neurosci, 2:467-504, 1979.
10. Bray D, Thomas C: Polymerized actin in fibroblasts and brain. J Mol Biol, 105:527-530.
11. Chang CM, Goldman RD: The localization of actin-like fibers in cultured neuroblastoma cells are revealed by heavy meromyosin binding. J Cell Biol, 57:867-874, 1973.
12. Isenberg G, Small JV: Filamentous actin, 100A filaments and microtubules in neuroblastoma cells. Cytobiology, 16:326-344, 1978.

34. Nagara H, Doi H, Iwaki T, et al: Intracytoplasmic inclusion of Hirano type in Purkinje cells. Clin Neuropathol, 5:131-133, 1986.
35. Yamamoto T, Hirano A: Hirano bodies in the perikaryon of the Purkinje cell in a case of Alzheimer's disease. Acta Neuropathol (Berl), 67:167-169, 1985,
36. Ho K-L, Allevato PA: Hirano body in an inflammatory cell of leptomeningeal vessel infected by fungus Paecilomyces. Acta Neuropathol (Berl), 71:159-162, 1986.
37. Okamato K, Hirai S, Hirano A: Hirano bodies in myelinated fibers of hepatic encephalopathy. Acta Neuropathol (Berl), 58:307-310, 1982.
38. Tomonaga M: Hirano body in extraocular muscle, Acta Neuropathol (Berl), 60:309-313, 1983.
39. Ogata J, Budzilovich GN, Cravioto H: A study of rod-like structures (Hirano bodies) in 240 normal and pathology brains. Acta Neuropathol (Berl), 21:61-67, 1972.
40. Tomanaga M, Yamanouchi H, Kameyama M, et al: Hirano bodies observed in the brain of the aged. Nippon Ronen Isakkai Zasshi, 12:13-17, 1975.
41. Gibson PH, Stones M, Tomlinson BE: Senile changes in human neocortex and hippocampus compared by the use of the electron and light microscopes. J Neurol Sci, 27:389-405, 1976.
42. Gibson PH, Tomlinson BE: Numbers of Hirano bodies in the hippocampus of normal and demented people with Alzheimer's disease. J Neurol Sci 33:199-206, 1977.
43. Schochet SS Jr, Lampert PW, Lindenberg R: Fine structure of the Pick and Hirano bodies in a case of Pick's disease. Acta Neuropathol (Berl), 11:330-337, 1968.
44. Anzil AP, Herrlinger H, Blinzinger K, et al: Ultrastructure of brain and nerve biopsy in Wilson disease. Arch Neurol, 31:94-100, 1974.
45. Cartier L, Galvez S, Gadjusek DC: Familial clustering of the ataxic form of Creutzfeld-Jakob disease. J Neurol Neurosurg Psychiatry, 48:234-238, 1985.
46. Atsumi T, Yamamura Y, Sato T, et al: Hirano bodies in the axon of peripheral nerves in a case with progressive external ophthalmoplegia with multisystem involvement. Acta Neuropathol (Berl), 49:95-100, 1980.
47. Ulrich J, Herschkowitz N: Seitelberger's connatal form of Pelizaeus-Mergbacher disease. Case report, clinical, pathological and biochemical findings. Acta Neuropathol (Berl) 40:129-136, 1977.
48. Beal JA: Morphogenesis of the Hirano body in neurons of the squirrel monkey dorsal horn. J Neurocytol, 7:395-403, 1978.
49. Jacobs JM, Cavanagh JB: Aggregations of filaments in Schwann cells of spinal roots of the normal rat. J Neurocytol, 1:161-167, 1973.
50. Spencer PS, Thomas PK: Ultrastructural studies of the dying-back process, II., The sequestration and removal by Schwann cells and oligodendrocytes of organelles from normal and diseased axons. J Neurocytol, 3:763-783, 1974.
51. Thomas PK, King RH, Sharma AK: Changes with age in the peripheral nerve of the rat. An ultrastructural study. Acta Neuropathol (Berl), 52:1-6, 1980.
52. Yagishita S, Itoh Y, Nakano T, et al: Crystalloid inclusions reminiscent of Hirano bodies in autolyzed peripheral nerve of normal Wistar rats. Acta Neuropathol (Berl), 47:231-236, 1979.
53. David-Ferreira JF, David-Ferreira KL, Gibbs CJ, Jr, et al: Scrapie in mouse: Ultrastructural observations in the cerebral cortex. Proc Soc Exp Biol, 127:313-320, 1968.
54. Sima AA, Hinton D: Hirano bodies in the distal symmetric polyneuropathy of the spontaneously diabetic BB-Wistar rat. Acta Neuro (Scand), 68:107-112, 1983.

13. Kuczmarski ER, Rosenbaum JL: Studies on the organization and localization of actin and myosin in neurons. J Cell Biol, 80:356-371, 1979.
14. Letourneau PC: Immunocytochemical evidence for colocalization in neurite growth cones of actin and myosin and their relationship to cell-substratum adhesion. Develop Biol, 85:113-122, 1981.
15. Letourneau PC, Ressler AH: Differences in the organization of actin in the growth cones compared with the neurites of cultured neurons from chick embryos. J Cell Biol, 97:963-973, 1983.
16. Luduena MA, Wessels NK: Cell locomotion, nerve elongation and microfilaments. Develop Biol, 30:427-440, 1973.
17. Spooner BS, Holladay CR: Distribution of tubulin and actin in neurites and growth cones of differentiating nerve cells. Cell Motility, 1:167-178, 1981.
18. Tsui HCT, Ris H, Klein WL: Ultrastructural networks in growth cones and neurites of cultured central nervous system neurons. Proc Natl Acad Sci U.S.A., 80:5779-5783, 1983.
19. Fifkova E, Delay RJ: Cytoplasmic actin in neuronal processes as a possible mediator of synaptic plasticity. J Cell Biol, 95:345-350, 1982.
20. Ishikawa H, Bischoff R, Holtzer H: Formation of arrowhead complexes with heavy meromyosin in a variety of cell types. J Cell Biol, 43:312-328, 1969,
21. Gambetti P, Schecket G, Ghetti B, et al: Neurofibrillary changes in human brain: an immunocytochemical study with a neurofilament antiserum. J Neuropathol Exp Neurol, 42:69-79, 1983.
22. Anderton BH, Calvert R, Probst A, et al: Antibody studies of neurofilaments and neurofibrillary tangles. J Submicrosc Cytol, 16:63-64, 1984.
23. Perry G, Rizzuto N, Autilio-Gambetti L, et al: Paired helical filaments from Alzheimer disease patients contain cytoskeletal components. Proc Natl Acad Sci U.S.A., 82:3916-3920, 1985.
24. Perry G, Selkoe DJ, Block BR, et al: Electron microscopic localization of Alzheimer neurofibrillary tangle components recognized by an antiserum to paired helical filaments. J Neuropath Exp Neurol, 45:161-168, 1986.
25. Perry G, Stewart D, Friedman R, et al: Filaments of Pick's bodies contain altered cytoskeletal elements. Am J Pathol, in press, 1987.
26. Hirano A, Malamud N, Elizan TS, et al: Amyotrophic lateral sclerosis and Parkinsonism-dementia complex on Guam. Arch Neurol: 15:35-51, 1966.
27. Hirano A, Dembitzer HM, Kurland LT, et al: The fine structure of some intraganglionic alterations: Neurofibrillary tangles, granulovacuolar bodies and "rod-like" structures as seen in Guam amyotrophic lateral sclerosis and Parkinsonism-dementia complex. J Neuropathol Exp Neurol, 27:167-182, 1968.
28. Goldman JE: The association of actin with Hirano bodies. J Neuropathol Exp Neurol, 42:146-152, 1983.
29. Schochet SS Jr, Hardman JM, Ladewig PP, et al: Intraneuronal conglomerates in sporadic motor neuron disease. Arch Neurol, 20:548-553, 1969.
30. Field EJ, Mathews JD and Raine CS: Electron microscopic observations on the cerebellar cortex in kuru. J Neurol Sci, 8:209-224, 1969.
31. Gibson PH: Light and electron microscopic observations on the relationship between Hirano bodies, neuron and glial perikarya in the human hippocampus. Acta Neuropathol (Berl), 30:165-171, 1978.
32. Ramsey HJ: Altered synaptic terminals in cortex near tumor. Am J Pathol, 51:1093-1109, 1967.
33. Hirano A, Dembitzer HM: Eosinophilic rod-like strulctures in myelinated fibers of hamster spinal roots. Neuropathol Appl Neurobiol, 2:225-232, 1976.

55. Brown HR, Jervis GA, Thormar H: Ultrastructural and histological studies of brains of ferrets inoculated with subacute sclerosing panencephalitis: similarities to human disease. J Neuropathol Exp Neurol, 36:653-655, 1977.
56. Goldman JE, Suzuki K: Immunocytochemical evidence for the associateion of actin with Hirano bodies, in: "IXth International Congress of Neuropatnology," M. Goldstein, ed: Springer-Verlag, New York, 1983.
57. Nagara H, Yakajima K, Suzuki K: An ultrastructural study of the cerebellum of the brindled mouse. Acta Neuropathol (Berl), 52:41-50, 1980.
58. Peterson C, Suzuki K, Kress Y, et al: Abnormalities of dendritic actin organization in the brindled mouse. Brain Res, 382:205-212, 1986.
59. Yajima K, Suzuki K: Neuronal degeneration in the brain of the brindled mouse. Acta Neuropathol (Berl), 45:17-25, 1979.
60. Craig SW, Pollard T: Actin-binding proteins. Trends Biochem Sci, 7:88-95, 1982.
61. Pollard TD, Aebi U, Cooper JA, et al: Actin structure, polymerization and gelation. Cold Spring Harbor Symp on Quant Biol, 46:513-524, 1981.
62. Stossel TP, Chaponnier C, Ezzell RM, et al: Nonmuscle actin binding proteins. Ann Rev Cell Biol, 1:353-402, 1985.
63. Weeds A: Actin-binding proteins-Regulators of cell architecture and motility. Nature (Lond), 296:811-815, 1982.
64. Galloway PG, Perry G, Gambetti P: Hirano body filaments contain actin and actin associated proteins. J Neuropathol Exp Neurol, 46:185-199, 1987.
65. Giometti CS, Anderson NL: Tropomyosin heterogeneity in human cells. J Biol Chem, 259:14113-14120, 1984.
66. Fine RE, Blitz AL, Hitchcock SE, et al: Tropomyosin in brain and growing neurones. Nature New Biol, 245:182-186, 1973.
67. Fine RE, Blitz AL: A chemical comparison of tropomyosins from muscle and nonmuscle tissues. J Mol Biol, 95:447-454, 1975.
68. Schliwa M: Proteins associated with cytoplasmic actin. Cell, 25:587-590, 1981.
69. Hitchcock-DeGregori SE, Lewis SF, Chou TM-T: Tropomyosin lysine reactivities and relationship to coiled-coil structure. Biochemistry, 24:3305-3314, 1985.
70. Joyce NC, Haire MF, Palade GE: Contractile proteins in pericytes. I. Immunoperoxidase localization of the tropomyosin. J Cell Biol, 100:1379-1386, 1985.
71. Warren RH, Gordon E, Azarina R: Tropomyosin in peripheral ruffles of cultured rat kidney cells. Eur J Cell Biol, 38:245-253, 1985.
72. Smillie LB: Structure and functions of tropomyosins from muscle and non-muscle sources. Trends Biochem Sci, 4:151-155, 1979.
73. Wegner A: Kinetic analysis of actin assembly suggests that tropomyosin inhibits spontaneous fragmentation of actin filaments. J Mol Biol, 161:217-227, 1982.
74. Bernstein BW, Bamburg JR: Tropomyosin binding to F-actin protects the F-actin from disassembly by actin de-polymerizing factor (ADF). Cell Motil, 2:1-8, 1982.
75. Fattoum A, Hartwig JH, Stossel HP: Isolation and some structural and function properties of macrophage tropomyosin. Biochemistry, 22:1187-1193, 1983.
76. Duhaiman AS, Bamburg JR: Isolation of brain alpha-actinin. Its characterization and a comparison of its properties with those of muscle alpha-actinin. Biochemistry, 23:1600-1608, 1984.
77. Bennett JP, Zaner KS, Stossel TP: Isolation and some properties of macrophage alpha-actinin: evidence that it is not an actin gelling protein. Biochemistry, 23:5081-5086, 1984.

78. Geiger B: A 130K protein from chicken gizzard: its localization at the termini of microfilament bundles in cultured chicken cells. Cell, 18:193-205, 1979.
79. Feramisco JR, Burridge K: A rapid purification of alpha-actinin, filamin and a 130,000-dalton protein from smooth muscle. J Biol Chem, 255:1194-1199, 1980.
80. Burridge K, Feramisco JR: Alpha-actinin and vinculin from non-muscle cells: calcium sensitive interactions with actin. Cold Spring Harbor Symp Quant Biol, 46:587-597, 1982.
81. Isenberg G, Leonard K, Jockusch BM: Structural aspects of vinculin-actin interaction. J Mol Biol, 158:231-249, 1982.
82. Jockusch BM, Isenberg G: Interactions of alpha-actinin and vinculin with actin: opposite effects on filament network formation. Proc Natl Acad Sci U.S.A., 78:3005-3009, 1981.
83. Jockuscn BM, Isenberg G: Vinculin and alpha-actinin: interaction with actin and effect on microfilament network formation. Cold Spring Harbor Symp on Quant Biol, 46:613-623, 1982.
84. Wilkins JA, Lin S: High affinity interaction of vinculin with actin filaments in vitro. Cell, 28:83-90, 1982.
85. Wilkins JA, Lin S: A re-examination of the interaction of vinculin and actin. J Cell Biol, 102:1085-1092, 1986.
86. Evans RR, Robson RM, Stromer MH: Properties of smooth muscle vinculin. J Biol Chem, 259:3916-3924, 1984.
87. Otto J: Vinculin: intereactions with actin, with itself, and with a 220,000 dalton vinculin binding protein. J Cell Biol, 97:282a,1983.
88. Otto JJ: The lack of interaction between vinculin and actin. Cell Motility and Cytoskeleton, 6:48-55, 1986.
89. Rosenfeld GA, Hou DC, Dingus J, et al: Isolation and purification of human platelet vinculin. J Cell Biol, 100:669-676, 1985.
90. O'Brien L, Shelly S, Towfighi J, et al: Crystalline ribosomes are present in brains from senile humans. Proc Natl Acad Sci U.S.A., 77:2260-2264, 1980.
91. Mori H, Tomanaga M, Baba N, et al: The structure analysis of Hirano bodies by digital processing of electron micrographs. Acta Neuropathol (Berl), 71:32-37, 1986.
92. Wulf E, Deboben A, Bautz FA, et al: Fluorescent phallotoxin, a tool for the visualization of cellular actin. Proc Natl Acad Sci U.S.A., 76:4498-4502, 1979.
93. Kidd M: Paired helical filaments in electron microscopy of Alzheimer's disease. Nature, 197:192-193, 1963.
94. Iqbal K, Grundke-Iqbal I, Wisniewski: Microtubule associated protein tau is a major constituent of Alzheimer paired helical filaments. J Biol Chem 261:6084-6089, 1986.
95. Kosik KS, Joachim CL, Selkoe DJ: Microtubule associated protein, tau, is a major antigenic component of paired helical filaments in Alzheimer disease. Proc Natl Acad Sci U.S.A., 83:4044-4049, 1986.
96. Nukina N, Ihara Y: One of the antigenic determinants of paired helical filaments is related to tau protein. J Biochem, 99:1541-1544, 1986.
97. Autilio-Gambetti L, Gambetti P, Crane R: Paired helical filaments: Relatedness to neurofilaments shown by silver staining and reactivity with monoclonal antibodies, in: R. Katzman, edi. "Banbury Report 15: Biological Aspects of Alzheimer Disease", Cold Spring Harbor Laboratory, Cold Spring Harbor, 1984.
98. Kosik KS, Duffy LK, Dowling MM, et al: Microtubule associated protein 2: Monoclonal antibodies demonstrate the selective incorporation of certain epitopes into Alzheimer neurofibrillary tangles. Proc Natl Acad Sci U.S.A., 81:7941-7945, 1984.
99. Brion JP, Passariero H, Nunez J, et al: Immunologic determinants of tau protein are present in neurofibrillary tangles of Alzheimer disease. Arch Biol, 95:229-235, 1985.

100. Wood JG, Mirra SS, Pollock NJ, et al: Neurofibrillary tangles of Alzheimer disease share antigenic determinants with the axonal microtubule-associated protein tau. Proc Natl Acad Sci U.S.A., 83:4040-4043.
101. Galloway PG, Perry G, Kosik K, et al: Hirano bodies contain tau protein. Brain Res, in press, 1987.
102. Drubin DG, Caput D, Kirschner MW: Studies on the expression of the microtubule associated protein tau, during mouse brain development, with newly isolated DNA probes. J Cell Biol, 98:1090-1097.
103. Cleveland DW, Hwo SY, Kirschner MW: Physical and chemical properties of purified tau factor and the role of tau in microtubule assembly. J Mol Biol, 116:227-247, 1984.
104. Griffith LM, Pollard TD: The interaction of actin filaments with microtubules and microtubule-associated proteins. J Biol Chem, 257:9143-9151, 1982.
105. DeCamilli PP, Miller P, Navone F, et al: Distribution of microtubule associated protein 2 (MAP2) in the nervous system of the rat studied by immunofluorescence. Neuroscience, 11:819-846, 1984.
106. Goldman JE: Cytoskeletal constituents of Hirano bodies. J Neuropathol Exp Neurol, 45:334a, 1986.

CONTRIBUTORS

Anderton, Brian H. Department of Immunology, St George's Hospital Medical School, Cranmer Terrace, London SW17 ORE, United Kingdom

Autilio-Gambetti, Lucila. Division of Neuropathology, Institute of Pathology, Case Western Reserve University, Cleveland, Ohio 44106

Baudier, Jacques. Department of Biochemistry, University of California, Berkeley, California 94720

Brion, Jean-Pierre. Department of Immunology, St George's Hospital Medical School, Cranmer Terrace, London SW17 ORE, United Kingdom

Castaño, Eduardo M. Department of Pathology, New York University Medical Center, 550 First Avenue, New York, New York 10016

Chin, Steven SM. Department of Pharmacology, New York University School of Medicine, New York, New York 10016

Cole, R David. Department of Biochemistry, University of California, Berkeley, California 94720

Crowe, Alex. Department of Pathology and The Rose F. Kennedy Center for Research in Mental Retardation and Human Development, Albert Einstein College of Medicine, Bronx, New York 10461

Deerinck, Thomas. Laboratory of Neurocytology, M-008, Department of Neurosciences, University of California, San Diego, La Jolla, California 92093

Dickson, Dennis W. Department of Pathology and The Rose F. Kennedy Center for Research in Mental Retardation and Human Development, Albert Einstein College of Medicine, Bronx, New York 10461

Ellisman, Mark. Laboratory of Neurocytology, M-008, Department of Neurosciences, University of California, San Diego, La Jolla, California 92093

Frangione, Blas. Kaplan Cancer Center and Department of Pathology, New York University Medical Center, 550 First Avenue New York, New York 10016

Galloway, Pamela. Department of Pathology, Cleveland Metropolitan General Hospital, Case Western Reserve University, Cleveland, Ohio 44109

Gambetti, Pierluigi. Division of Neuropathology, Institute of Pathology, Case Western Reserve University, Cleveland, Ohio 44106

Gauthier, S. McGill Center for the Study of Ageing, Montreal General Hospital, Montreal, Quebec, Canada H3G 1A4

Goldman, James E. Department of Pathology, Albert Einstein College of Medicine, Bronx, New York 10461

Grundke-Iqbal, Inge. Institute for Basic Research in Developmental Disabilities, 1050 Forest Hill Road, Staten Island, New York 10314

Houghton, S. Marine Biological Laboratory, Woods Hole, Massachusetts 02543

Hue, DP. Department of Immunology, St George's Hospital Medical School, Cranmer Terrace, London SW17 0RE, United Kingdom

Iqbal, Khalid. Institute for Basic Research in Developmental Disabilities, 1050 Forest Hill Road, Staten Island, New York 10314

Kosik, Kenneth S. Center for Neurologic Diseases, Brigham and Women's Hospital, Department of Medicine, Harvard Medical School, Boston, Massachusetts 02115

Ksiezak-Reding, Hanna. Department of Pathology and The Rose F. Kennedy Center for Research in Mental Retardation and Human Development, Albert Einstein College of Medicine, Bronx, New York 10461

Leblanc, R. Montreal Neurological Institute, Montreal, Quebec, Canada H3A 2B4

Liem, Ronald KH. Department of Pharmacology, New York University School of Medicine, New York, New York 10016

Manetto, Valeria. Division of Neuropathology, Institute of Pathology, Case Western Reserve University, Cleveland, Ohio 44106

Metuzals, Janis. Department of Pathology, Faculty of Health Sciences, University of Ottawa, Ottawa, Ontario, Canada K1H 8M5

Mirra, Suzanne S. Veteran's Administration Medical Center, Emory University School of Medicine, Atlanta, Georgia 30033

Mulvihill, Paul. Division of Neuropathology, Institute of Pathology, Case Western Reserve University, Cleveland, Ohio 44106

Neve, Rachael. Neurosciences Department, Children's Hospital Medical Center, Harvard Medical School, Boston, Massachusetts 02115

Onorato, Michelle. Division of Neuropathology, Institute of Pathology, Case Western Reserve University, Cleveland, Ohio 44106

Perry, George. Division of Neuropathology, Institute of Pathology, Case Western Reserve University, Cleveland, Ohio 44106

Pollock, Nancy J. Department of Anatomy and Cell Biology, Emory University School of Medicine, Atlanta, Georgia 30332

Ranganathan, Rama. Laboratory of Neurocytology, M-008, Department of Neurosciences, University of California, San Diego, La Jolla, California 92093

Robitaille, Y. Montreal Neurological Institute, Montreal, Quebec, Canada H3A 2B4

Robinson, PA. Department of Immunology, St George's Hospital Medical School, Cranmer Terrace, London SW17 ORE, United Kingdom

Selkoe, Dennis J. Departments of Neurology and Neuroscience, Harvard Medical School and Center for Neurologic Diseases, Brigham and Women's Hospital, Boston, Massachusetts 02115

Tabaton, Massimo. Division of Neuropathology, Institute of Pathology, Case Western Reserve University, Cleveland, Ohio 44106

Terry, Robert. Department of Neurosciences, M-022, University of California, San Diego, La Jolla, California 92093

Weiss, Madelyn. Division of Neuropathology, Institute of Pathology, Case Western Reserve University, Cleveland, Ohio 44106

Wisniewski, Henryk M. Institute for Basic Research in Developmental Disabilities, 1050 Forest Hill Road, Staten Island, New York 100314

Wood, John G. Department of Anatomy and Cell Biology, Emory University School of Medicine, Atlanta, Georgia 30332

Yen, Shu-Hui. Department of Pathology and the Rose F. Kennedy Center for Research in Mental Retardation and Human Development, Albert Einstein College of Medicine, Bronx, New York 10461

Young, Stephan. Laboratory of Neurocytology, M-008, Department of Neurosciences, University of California, San Diego, La Jolla, California 92093

[illegible], [illegible]. [illegible], [illegible], [illegible], [illegible].

[illegible], [illegible]. [illegible] [illegible] [illegible], [illegible].

[illegible], [illegible]. [illegible] of [illegible] in [illegible], [illegible], [illegible].

[illegible] [illegible], [illegible], [illegible].

[illegible] [illegible], [illegible], [illegible].

[illegible]. [illegible] and [illegible], [illegible] [illegible], [illegible].

INDEX

GPSR Compliance

The European Union's (EU) General Product Safety Regulation (GPSR) is a set of rules that requires consumer products to be safe and our obligations to ensure this.

If you have any concerns about our products, you can contact us on ProductSafety@springernature.com

In case Publisher is established outside the EU, the EU authorized representative is:

Springer Nature Customer Service Center GmbH
Europaplatz 3
69115 Heidelberg, Germany

Batch number: 10682325

Printed by Printforce, the Netherlands